Green Your Home

The Complete Guide to Making Your New or Existing Home Environmentally Healthy

By Jeanne Roberts

Green Your Home

THE COMPLETE GUIDE TO MAKING YOUR NEW OR EXISTING HOME ENVIRONMENTALLY HEALTHY

1405 SW 6th Ave • Ocala, Florida 34471 • 800-814-1132 • 352-622-1875–Fax
Web site: www.atlantic-pub.com • E-mail: sales@atlantic-pub.com
SAN Number: 268-1250

ISBN-13: 978-1-60138-128-6 ISBN-10: 1-60138-128-X

Library of Congress Cataloging-in-Publication Data

Roberts, Jeanne, 1944-
Green your home: the complete guide to making your new or existing home environmentally healthy/by Jeanne Roberts.
p. cm.
Includes bibliographical references and index.
ISBN-13: 978-1-60138-128-6 (alk. paper)
ISBN-10: 1-60138-128-X (alk. paper)
1. Ecological houses. 2. Sustainable living. 3. Environmental protection--Citizen participation. 4. Home economics. I. Title.

TH4860.R63 2008
643--dc22

2008010970

INTERIOR LAYOUT DESIGN: Vickie Taylor • vtaylor@atlantic-pub.com

Printed in the United States

Printed on Recycled Paper

Author Dedication

This book is dedicated to all the wonderful people who made it happen, including those who submitted Case Studies, the editor who walked me through it, the proofreaders who were so exactingly correct, and finally my children, Laura, Deborah, Jesse, and Josh, who — grown up — have turned out to be much better people than I deserve.

We recently lost our beloved pet "Bear," who was not only our best and dearest friend but also the "Vice President of Sunshine" here at Atlantic Publishing. He did not receive a salary but worked tirelessly 24 hours a day to please his parents. Bear was a rescue dog that turned around and showered myself, my wife Sherri, his grandparents Jean, Bob and Nancy and every person and animal he met (maybe not rabbits) with friendship and love. He made a lot of people smile every day.

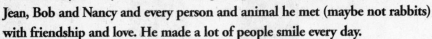

We wanted you to know that a portion of the profits of this book will be donated to The Humane Society of the United States. ***–Douglas & Sherri Brown***

The human-animal bond is as old as human history. We cherish our animal companions for their unconditional affection and acceptance. We feel a thrill when we glimpse wild creatures in their natural habitat or in our own backyard.

Unfortunately, the human-animal bond has at times been weakened. Humans have exploited some animal species to the point of extinction.

The Humane Society of the United States makes a difference in the lives of animals here at home and worldwide. The HSUS is dedicated to creating a world where our relationship with animals is guided by compassion. We seek a truly humane society in which animals are respected for their intrinsic value, and where the human-animal bond is strong.

Want to help animals? We have plenty of suggestions. Adopt a pet from a local shelter, join The Humane Society and be a part of our work to help companion animals and wildlife. You will be funding our educational, legislative, investigative and outreach projects in the U.S. and across the globe.

Or perhaps you'd like to make a memorial donation in honor of a pet, friend or relative? You can through our Kindred Spirits program. And if you'd like to contribute in a more structured way, our Planned Giving Office has suggestions about estate planning, annuities, and even gifts of stock that avoid capital gains taxes.

Maybe you have land that you would like to preserve as a lasting habitat for wildlife. Our Wildlife Land Trust can help you. Perhaps the land you want to share is a backyard—that's enough. Our Urban Wildlife Sanctuary Program will show you how to create a habitat for your wild neighbors.

So you see, it's easy to help animals. And The HSUS is here to help.

THE HUMANE SOCIETY OF THE UNITED STATES®

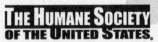

2100 L Street NW • Washington, DC 20037 • 202-452-1100

www.hsus.org

Contents

Foreword

By Kelly Hart

Creator and Webmaster of **www.greenhomebuilding.com**

Jeanne Roberts has clearly done her homework in providing convincing data and suggestions for ways to conserve electricity and water, ways to create a healthy home environment, and ways to lessen our adverse physical impact while we live our lives.

It is crucial for all of us to find a way to alleviate the burden that mankind has placed on the ecological balance of life on earth. Exacerbated by our increasing numbers, we have been consuming more material resources every year for centuries, and most of these resources are finite and irreplaceable.

We have become dependent on a pattern of energy usage that is not sustainable. The infrastructure for our lifestyle was established without thought for the consequences, but now it is abundantly clear that our gluttonous appetite for fossil fuel will cost us dearly. We are paying for this not only with escalating fuel prices, but with the need to endure increasing pollution of the air we breathe, the water we drink, the soil we till, and the oceans we harvest. As if this were not bad enough, now we have the very real experience of global warming to deal with, as it levies its toll of extreme weather, rising sea levels, and erratic climatic conditions.

It is imperative that each of us now takes responsibility for changing our course to find a sustainable way to live. This is largely a matter of living with the conscious intention to conserve the energy and water we use, and the materials we utilize. We need to find ways to consume less.

The good news is that this new lifestyle does not necessarily mean deprivation or dissatisfaction; we can still enjoy technical innovations and the comforts of a modern life. Through a blend of taking advantage of new, more efficient appliances, and applying simple concepts for designing and building our homes, we can find the cutting-edge way toward sustainability.

I can tell you from experience that living in a well-designed passive solar home that heats itself and provides much of its own electricity and hot water from solar panels cannot be beat for the comfort that it provides. I am speaking of both the temperate interior climate and in the peace of mind that I am not relying on fossil fuel for my needs and that I am doing my part to consume less.

Jeanne Roberts has assembled here a very useful guide to help homeowners navigate the delicate and necessary task of providing a healthy home in such a way as to assure that future generations may also enjoy healthy homes. At this time in our technological society this is not an easy task, since there is a bewildering array of choices that we are presented with, and no clear sign posts for which choices will lead us in the right direction.

When it comes time to either remodel an existing home or build a new one, she explains what some of your better choices might be for all of those specific questions that arise about roofs, windows, heating systems, plumbing fixtures, lighting, paints, etc. Guidance is also provided for basic building technologies, both traditional, natural options, as well as some of the newer, more industrial alternatives. The merits and demerits of these various systems are discussed.

"Case Studies" that profile the work of a variety of informants regarding relevant topics are an interesting and informative addition to this book. These provide real insight into the tangible use of what are often hypothetical or theoretical issues.

Wise use of this guide will help you become a responsible citizen of planet earth and help preserve a healthy environment for our descendants.

www.greenhomebuilding.com
www.dreamgreenhomes.com
www.earthbagbuilding.com

Introduction: The Meaning of Green

Making your home "green" means making it environmentally friendly. A green home is good for you and the planet. The result is sustainability. Sustainability is the science and art of occupying the planet in such a way that the needs of the current inhabitants are met without compromising the needs of future inhabitants.

Unfortunately, sustainability implies limits. Half the power produced in this country is generated by burning coal, another quarter by burning gas and oil. In 1949, the United States used about three quads of energy. A quad is the equivalent of one quadrillion British thermal units (Btus). A Btu is the amount of energy needed to raise the temperature of a pound of water by one degree Fahrenheit. In 2005, the United States used almost 101 quads, and the most recent figures indicate energy use will rise to 102 quads in 2007.

In 1997, transportation accounted for 27 percent of energy consumption, industrial use for another 37 percent. Residential and commercial facilities accounted for 36 percent. More than one-third of energy consumption keeps us warm or cool and lights our way. Our cars use 385 million gallons of gas per day, or 69 percent of all the fuel used in transportation.

Where will the energy come from? Everyone assumes it will come from oil or natural gas, but current United States Geological Survey (USG) figures indicate there are only about 1.7 trillion barrels of conventional oil remaining on earth and an equal amount of natural gas. This is an optimistic but flawed estimate. The Energy Information Administration (EIA) expects oil and gas production to grow almost 2 percent per year from 2001 to 2025. At that growth rate, the life of oil or gas as a resource diminishes to 41 years. Using United States Geological Survey figures as a guide, we can expect oil and gas production to peak in 2015, just seven years from now. Peak oil is the time at which maximum petroleum production occurs, after which production can be expected to decline. Unfortunately, declining production will not keep pace with the growing need.

Energy companies that converted coal-burning plants to natural gas at a cost of billions of dollars soon will have to convert back. Even then, reserves of coal in this country will last only until about 2080. What happens after that?

As we approach this impasse, we also have to look back at the effects of our consumption. Forty percent of carbon dioxide emissions come from the burning of fossil fuels, such as oil, coal, and natural gas, to generate electrical power to power homes and businesses. As global emissions of carbon dioxide rise, the planet warms. The average atmospheric concentration of carbon dioxide reached 380 parts per million in 2005. The Intergovernmental Panel on Climate Change (IPCC), a global body of roughly 2,000 scientists, estimates that the current concentration of carbon dioxide in the atmosphere has not been exceeded over the last 420,000 years and probably not during the past 20 million years. In 2006, the World Meteorological Organization, in its annual greenhouse gas bulletin based on samples from 40 countries, reported that levels of carbon dioxide in the air had reached 381.2 parts per million, a level not seen in 650,000 years. Over the past 30 years, global temperatures have risen by 0.6 degrees Celsius, or 1.08 degrees Fahrenheit. The Intergovernmental Panel on Climate Change projects that this warming trend will continue as carbon dioxide emissions become concentrated and that global temperatures will rise by a minimum of 1.4 degrees Celsius. Some scientists, fearing the worst, estimate the possible rise at 5.8 degrees Celsius, or 10.44 degrees Fahrenheit, by 2100. A rise of this magnitude would make some areas of earth uninhabitable.

The effects of global warming have been documented. The World Health Organization says 150,000 deaths each year can be attributed to the effects of climate change. Sea levels have risen about six inches over the past century, while hurricanes have become increasingly severe, as demonstrated by Hurricane Katrina in 2005, which devastated portions of New Orleans. Rising sea levels and more severe storms will threaten the homes and the livelihoods of millions of Americans living in the Gulf of Mexico area or along the Atlantic Seaboard.

By 2100, the Intergovernmental Panel on Climate Change estimates sea levels will rise from 7 to 23 inches. This estimate is based on poorly understood phenomena; the earth, during the time humans have occupied it, has never warmed as fast as it is warming now. Greenland and Antarctica, the two polar landmasses, together make up almost 69 percent of the earth's fresh water, locked up in ice, and their rate of melting is speeding up. If these thawed quickly and simultaneously, sea levels would rise 215 feet. This does not even take into account glacial melting from the Alps, the Andes, and the Himalayas, which has also begun to occur and more rapidly than scientific calculations had predicted.

"The current dynamical changes that we are seeing on the ice sheet are not captured in any climate model," said Prasad Gogineni, director of the Center for Remote Sensing of Ice Sheets at the University of Kansas. "That seems to indicate a huge uncertainty."

Some scientists are concerned that Intergovernmental Panel on Climate Change predictions might err on the side of caution. Greenland's melting ice is contributing more than half a millimeter per year — twice the panel's 2002 estimate. Many scientists agree that the only way to prevent catastrophic warming is to cut emissions by about 70 percent before 2050. Unfortunately, much of the growth in carbon emissions is expected to come from such developing countries as China and India, and both have already shown their unwillingness to curb their economic growth to benefit developed countries, such as the United States.

The Intergovernmental Panel on Climate Change issued a final report in November 2007, concluding that global warming is now "unequivocal." In other words, there is no longer any doubt that earth's climate is changing for the worse as a result of our bad habits, and future climactic events, such as hurricanes, floods, droughts, loss of crops, and diminished or obliterated habitat will occur suddenly, dramatically, and with no hope of mitigation, even if immediate steps are taken to curb greenhouse gases.

In human terms, warming means an increase in air pollution as harmful gases are trapped at or near ground level. One significant development of this pollution is an increase in asthma. In 2002, about 30.8 million people in this country had asthma, 9 million of them children. Childhood rates of asthma have doubled since 1980, and this illness is the primary cause of hospitalization in children as well as the most widespread chronic disease among children. Asthma causes more school absences than any other health concern. Asthma also is on the rise among women, and in 2001, 65 percent of asthma-related deaths were female. Another segment of the population, the urban poor, has seen skyrocketing rates of asthma as our largest cities become too polluted to support a healthy life style.

Global warming, due to burning fossil fuels, also means depletion of the ozone layer. This is a two-fold threat: germs breed and travel better at warmer temperatures, and increases in ultraviolet (UV) radiation due to ozone depletion harm our immune systems, making us less able to fight off increasingly prevalent and antibiotic-resistant diseases and giving rise to more autoimmune diseases. Ultraviolet radiation also causes skin cancer and genetic modifications at the most basic, cellular level of the body. In humans, these mutations might take generations to display. Among mammalian and invertebrate populations, the effects are already evident: amphibian populations are on the decline, according to the article by Adrienne Howse, "Where Have All the Frogs Gone," published at **ActionBioscience.org**. Scientists regard amphibians as natural indicators of our planet's health, in much the same way coal miners once used canaries to identify bad air in the mining tunnels.

The warning is clear: Unless we begin to conserve and reuse our resources and invest in alternative sources of energy — maybe under government sponsorship — our future, and the future of our children, will be dire.

What it Means to be Green

Making your home green can be as simple and inexpensive as installing a displacement device in your toilet or as elaborate and costly as replacing your windows with energy-efficient windows or insulating your home using a new, eco-friendly soy-foam insulation. Between these extremes, you can do any number of things to improve both your own health and that of the planet. Most projects cost between $1 and $2,500. Many larger greening projects can be specially financed through your utility company, a state agency, an organization such as Energy Smart, or even your bank. These loans offer low interest rates, flexible repayment terms, few or no fees, and provide you, the homeowners, with "green" credits to use to defray the capital gain if you sell your home. If you have a contractor, check with him or her for the latest updates on these "energy smart" loans.

There are many things you can do to increase your home or condo's sustainability factor. You can catch and store rainwater from your roof, xeriscape your yard or garden to reduce water usage, replace your toxic carpet with salvaged wood, put film on your windows and seal your doors, reduce the amount of trash you throw out, and buy cleaning products, paints, and stains that do not contribute to the earth's current state of toxic overload. These are just a few suggestions.

In the following chapters, I will show you how to use your washer and dryer, hot water heater, air conditioning, and other appliances to reduce your energy usage and your bill. I also will show which residential energy-saving programs have the biggest payoff in terms of money and sustainability. I will advise you on selecting building or remodeling supplies that do not further affect an already troubled planet and provide advice and suggestions from experts on such diverse topics as collecting rain from your roof and installing solar collectors. Finally, I will suggest ways in which you can identify and hire a green architect or builder if you want to rebuild in an entirely "earth-friendly" fashion. You can even join a grass-roots green organization or start your own, as hundreds of concerned citizens around the country already have. You can focus on water, waste, energy consumption, or any of the other areas in which growing populations have begun to put natural resources on the endangered list.

Once you begin, you will find more ways to reduce your impact on the environment — also known as your carbon footprint or your eco-footprint. You will audit your house, just the way the IRS audits you, to discover how much electricity you are using, or you can call your local electric utility to audit your home for you. You can buy a certain portion of your electricity as green power and buy renewable energy credits to offset your use. You can invest in solar-powered chargers for your electronics and run everything on power strips that can be shut off when not in use. You will begin to understand that achieving sustainable development — the ability of people to live on the planet without destroying it — is both a process and a goal.

We should do these things because we have reached a tipping point on the planet, where more people are consuming more resources. We cannot rightfully tell our children — or other developing nations — that they do not deserve to live as well as we have. All we can do now, at the beginning of the 21st century, is conserve what we have left so that remaining resources are more equitably distributed and continue to develop new, earth-friendly resources to take us into our future.

In case you still doubt the peril our earth is in and question sustainability as a necessary and viable solution, here are some interesting facts:

- The United States generates about 208 million tons of municipal solid waste per year, according to the National Institutes of Health. That is more than four pounds per person per day. Every little bit helps; recycling just one glass bottle conserves enough electricity to light a 100-watt bulb for four hours.

- According to the Energy Star Web site, if just one in ten homes used energy-efficient appliances, it would be equivalent to planting 1.7 million acres of new trees.

- Using a broom instead of the garden hose to clean your driveway can save 80 gallons of water, and turning the water off while you brush your teeth will save 4.5 gallons every time.

- Each person will receive almost 560 pieces of junk mail this year, which adds up nationally to 4.5 million tons, according to the Native Forest Network. About 44 percent of all junk mail is thrown in the trash, unopened and unread, and ends up in a landfill.

- Each year, Americans throw away 100 billion polyethylene plastic bags — from grocery and trash bags to those ultra-convenient sandwich bags. Unfortunately, plastics are made from petroleum

— the processing and burning of which is considered one of the main contributors to global warming, according to the Environmental Protection Agency (EPA).

- American households use 80 million pounds of pesticides each year, according to the EPA.

How Green Are You?

A carbon footprint is, essentially, the amount of energy and materials you, or your entire family, consume to maintain your life style. Your footprint is measured by your geographic location, the square footage of your apartment or house, the number of bedrooms, and the number of people who occupy the unit. Footprinting also may measure your water usage, the number of vehicles you own, what kinds of food you eat, and whether you fly.

No method of footprinting is truly comprehensive because the most energy-consumptive units, such as computers and televisions, are not counted. Nor are such amenities as pools, spas, and hot tubs calculated. Neither does the amount of trash you recycle, as opposed to the amount you simply send to a landfill, enter the equation, and disposing of trash in landfills is a significant portion of everyone's footprint. In fact, these assessments do not even allow for green improvements such as rooftop solar, ground source heat pumps, low-flow toilets, other green building modifications, or even for the fact that you might be opting in to your utility's renewable energy resources, such as wind and solar. My advice is to take the test, increase the result by 25 percent to allow for electronics, and deduct the same amount if you use energy-efficient lighting, recycle your cans, turn down your water heater, and keep your thermostat at 68 degrees in the winter and 78 in the summer.

Yahoo! has a relatively simple calculator. Earth Day has another, more comprehensive site. Clean Air, Cool Planet has two. There are others in abundance, from TreeHugger to Greenpeace, and I encourage you to find one you like and measure your carbon footprint. The results will surprise you. Go to these Web sites:

- **green.yahoo.com/calculator**
- **ecofoot.org**
- **www.safeclimate.org/calculator**
- **www.gmpvt.com/energy101/myhome.shtml**

Cheap & Easy

In this chapter, we will cover relatively inexpensive ways to reduce electricity and water consumption and the amount of trash we throw into landfills. We also will explore some of the new, eco-friendly lines of cleaning supplies, paints, stains, and remodeling products.

Electricity: Incandescents versus Fluorescents and LEDs

Incandescent light bulbs — the rounded ones you probably use in your lamps and overhead lighting fixtures — produce heat first, then give off light as a secondary by-product. First invented in 1879 by Thomas Edison, they were a technological advancement on par with space flight. Now they are obsolete energy hogs, using 70 percent more energy than their modern cousins and lasting a tenth as long.

You can change them, and your carbon footprint, simply by switching to modern, compact fluorescent bulbs. These bulbs produce 75 percent less heat, reducing the potential for accidental fires, and last ten times as long. They come in all shapes and sizes, from the familiar, rounded style to a corkscrew style, a two- or three-tube model, and even a mushroom-shaped bulb. They can produce everything from the soft, white light you are familiar with to a sharp, clear light more suitable for work areas, and they also come in colors. They cost from $10 to $17 for a three-pack, save $30 over their lifetime, and burn for 12,000 hours, compared to an incandescent bulb's 1,000 hours. If you multiply the cost of a standard light bulb by ten, the cost of the modern bulb is entirely in line. More important, according to the United States Department of Energy, if every house in the country replaced just one incandescent bulb with a compact fluorescent bulb, we could save enough energy to light more than 3 million homes for an entire year and eliminate the amount of greenhouse gas caused by 800,000 cars.

There also are light-emitting diodes, or LEDs. These newcomers on the lighting scene may be bluish-white, like automobile headlights, which makes them suitable for work areas but not necessarily for living areas. Warmer tones in light-emitting diodes are not currently much more efficient than incandescents, but experts predict that by 2015 — with significant advances in color-modulating and fixture retrofitting already under way — light-emitting diodes will dominate the lighting market in cost, efficiency, and color values. Light-emitting diodes manufactured now are sensitive to temperature, which reduces their energy efficiency. Light-emitting diodes also require a ballast. However, LEDs are more directional than incandescents or fluorescents, so they focus light better. Where applicable, as in task lighting, they provide 300 percent more focused lumens in half the wattage of an incandescent bulb and burn up to 40 times longer.

Forcefield, an online energy Web site, offers specific recommendations for fluorescents and LEDs. For greatest efficiency, use fluorescents of any type in work spaces where brightness is desired, including basements, garages, and that neglected area over the kitchen sink; use compact fluorescents in living spaces for a gentler ambience; halogen lighting works well in outdoor applications, where fluorescents sometimes fail due to temperature variants; and light-emitting diodes should be used for task lighting, nightlights, flashlights, and to light walks and pathways. For more information, go to: **www.otherpower.com/otherpower_lighting.html.**

Before you invest in new light bulbs, you should note the ways in which lighting efficiency is presented. Some LED manufacturers show only lumens per watt of output but do not figure in power wasted as heat, which is an important consideration for off-grid homes — homes not connected by power lines to a utility company. Others show lumens used by the light-emitting diode, which is the actual efficiency rating.

CASE STUDY: ENERGY-EFFICIENT LIGHTING

Lighting an Off-Grid Home

By Dan Fink

Homeowners choose to install energy-efficient devices for a variety of reasons, including lowering their electricity bills, reducing their carbon footprint, and in general being kind to the planet. But for that small group of people who live entirely off the grid, energy efficiency in all parts of the home is essential and has nothing to do with any "warm and fuzzy" feelings about saving the earth.

I have lived off the grid — 11 miles from the nearest power line — for the past 17 years. I designed my home from the ground up to be energy efficient, and lighting is a big part of that. I use only about 100 kilowatt-hours (kWh) per month to power my home, while the national average electricity usage in the United States is about 750 kilowatt-hours per month. According to the U.S. Energy Information Administration (EIA), lighting loads account for almost 9 percent of the average U.S. homeowner's electricity consumption. Efficient lighting becomes even more important with a solar electric power system such as mine. During the winter months at my 40 degrees latitude, incoming sun hours are reduced while indoor lighting hours increase due to longer nights and shorter days.

Grid electricity that you buy from the local utility is actually very cheap compared to producing your own with solar or wind power. I know that may sound strange to a city dweller whose electric bills are increasing each month, but becoming your own utility is very expensive. A good rough estimate is that a solar power system costs $2,000 per kilowatt-hour per day capacity and $3,000 per kilowatt-hour per day with

CASE STUDY: ENERGY-EFFICIENT LIGHTING

battery backup, according to Rex Ewing and Doug Pratt in their book *Got Sun, Go Solar* (Pixyjack Press, 2005). Considering that the average U.S. home uses 24 kilowatt-hours per day, that puts the cost of installing the equipment to provide all your own energy at $50,000 to $70,000. Because grid electricity from the utility costs on average only $.10 per kilowatt-hour in the United States, it would take a long time to pay back such an investment without government incentives.

My power system produces (and I use) on average 3.6 kilowatt-hours per day, and the system cost me about $3,000 per kilowatt-hour per day capacity, closely following the estimate above from Ewing and Pratt. In my off-grid home, every light bulb is a compact fluorescent (CF), no exceptions. Compact fluorescents provide the same lighting level as incandescents with 75 percent less waste of electricity as heat, according to the U.S. Department of Energy. My measured energy savings using CF bulbs have closely followed that figure. I use on average only 0.317 kilowatt-hours per day for lighting. If all my lights were incandescent, that would mean 1.267 kilowatt-hours per day for lighting — at an extra cost of $1,000 investment in the initial system, almost 10 percent of its cost. In an on-grid situation, the costs are not nearly as painful. The average U.S. homeowner spends $6.60 a month ($79 a year) for lighting and could reduce that to $1.65 a month ($20 a year) by installing efficient lighting. But that is assuming that electricity stays at $.10 a kilowatt-hour. It is already three times more expensive than that throughout most of Europe and is steadily growing in cost here in the United States. That $79 per year for lighting could easily balloon to far more in the near future.

You will notice that I do not mention LED lighting; I have CF lights only. That is because the LED technology that is on the market right now for lighting homes and other large areas is not nearly as efficient as fluorescent lighting of any type. That fact is changing quickly, however. LED manufacturers have new designs that are approaching CF bulbs in efficiency — but those new LEDs are still in the laboratory and are a few years from being on the market in the large arrays needed for lighting an entire room at a reasonable cost.

I prefer to look at my lighting situation from a very practical standpoint, because I do not have the option of buying more electricity from a utility if I use too much. My efficient CF light bulbs save me about one kilowatt-hour each day. Instead of wasting that lone kilowatt-hour on heat from inefficient light bulbs, I can use it to: watch three movies on my (energy-efficient) big-screen LCD television; work at my job via satellite Internet for five hours (ten hours if I use a laptop computer); or even run four loads of dishes through my dishwasher.

A popular joke among renewable energy enthusiasts is, "How can you tell which of your party guests lives off the grid?" Answer: "It is the person who roams around the host's house, turning off the lights and televisions in the rooms that are unoccupied." When you have to produce every watt-hour that you use, habits (such as turning off

CASE STUDY: ENERGY-EFFICIENT LIGHTING

lights and such when they are not needed) are very important — but efficient lighting makes them somewhat less urgent. A 100-watt incandescent light accidentally left on overnight at my house would mean no movie watching at all the next evening, while a 25-watt CF bulb (that provides the same light level) left on would mean only a minor inconvenience — an hour less spent on the computer the next day perhaps.

In conclusion, I would like to mention a very strange efficient-lighting tale from off the grid. My neighbor up the road had to spend a winter morning in 2006 frantically removing every efficient CF bulb he had and replacing all of them with the most inefficient, wasteful, energy-sucking incandescent bulbs he could lay his hands on! Why? We had received five feet of snow in 24 hours and the next day dawned both sunny and windy — but his low-budget solar and wind power system controls were all located a few hundred feet from the house, in the "power shed." He had no way to shut down either the solar panels or the wind turbine without digging a path through the chest-deep snow, an all-day proposition. The only solution was to use more electricity, as much as he possibly could, to avoid overcharging and damaging his battery bank. So he spent the day with all his extremely inefficient lights on, the TV and stereo blaring, using as much power as he could until the sun went down and the wind calmed.

Rest assured — as soon as he dug a path to his power shed and got the situation under control, the compact fluorescent bulbs went right back in the sockets. The old energy-hog incandescents were relegated once again to gathering dust in the back of a kitchen cabinet — right where they belong. Such is life off the grid, high in the mountains.

Dan Fink is the Technical director of Otherpower: a division of Forcefield. For more information on this company and its products, go to **www.otherpower.com/otherpower_lighting.html.**

Dan Fink's house at night, courtesy of Dan Fink, Otherpower

Going Green and Electric Appliances

Chances are you have some standard appliances, such as a gas or electric stove with or without a hood fan, a microwave, and a refrigerator. You also might have a toaster, a coffeepot, a toaster oven, a Crock-Pot or slow cooker, and possibly a blender or food processor. There are many small appliances available, but these are the most common.

You might not have a washer and dryer, but you probably use them in an apartment laundry room or laundromat; if you use them, they are part of your carbon footprint.

You most likely will have one television, possibly more, and also might have a compact disc player, a DVD player, or a game console that hooks up to your television. You likely have a computer, possibly even more than one. Elsewhere in your home or apartment, you could have an electric space heater, a window air conditioner, a radio, or a stereo with speakers. You probably own a vacuum cleaner.

You might be lucky enough to have a boat de-icer, an air compressor, an exercise machine, a trash compactor, a deep fryer, a dishwasher, a freezer, a hot tub, a swimming pool, or a heated waterbed.

If you have even half these electrical appliances, your footprint is larger than you think. To calculate it exactly, go to your local gas or electric utility's Web site and use its calculator, or go to the Department of Energy's appliance calculator Web page, at **www.energy.qld.gov.au/energywise_calculator**.

The average American household uses between 750 and 1,000 kilowatt-hours of electricity per month. It takes one-half ton of coal, or 1,000 pounds, to make 1,250 kilowatt-hours of electricity. This works out to about 53 pounds a day, or about 10,000 pounds per year, for a household using 800 kilowatt-hours.

If coal is your electric company's primary power resource, your family is adding 62 pounds of sulfur dioxide, 40 pounds of nitrogen oxide, and about 19,000 pounds of carbon dioxide to the air each year. If your electric company is using hydropower, it needs only 800 gallons of water to make 1,000 kilowatt-hours of electricity, and there are virtually zero emissions; the typical family uses about ten times that much water per month for other purposes.

Let us examine where your household uses that 1,000 kilowatt-hours of electricity per month. The average refrigerator uses 91 kilowatt-hours per month; the average computer uses 32. Your electric clothes dryer uses a whopping 120 kilowatt-hours, and gas is not that much more efficient; it merely uses a different resource. A dehumidifier uses about 200 kilowatt-hours, as does a humidifier, and that fancy new boat de-icer can use up to 720 kilowatt-hours. If you live in a cold climate, the plug-in heater on your car is using 180 kilowatt-hours, as is your pool, if you have one. Your water heater is using about 300 kilowatt-hours, but your heated waterbed is using more than half that. Your forced-air furnace is using only about 200 kilowatt-hours, but your hot tub is using almost 600. For what it is costing you, you could almost afford to move to Florida. Your utility company charges you by the kilowatt-hour. The average household uses about 900 kilowatt-hours per month. Your electric company charges you a cheaper, base rate for initial use — depending on your region and your utility company — and more for additional hours. It also charges more in the summertime, when air conditioning is needed. On average, it charges about $.06 for the first 40 kilowatt-hours and $.10 for every kilowatt-hour after that.

According to the Energy Star Web site, the average homeowner spends about $1,900 on energy in a year. Simply by switching to energy-efficient appliances that wear the Energy Star label, you could save $80 of that cost while reducing your carbon footprint. Energy-efficient appliances, while priced about the same as standard appliances, use 10 to 50 percent less energy and water to operate, meaning that over time, you recover the cost of the appliance itself. More important, the "lights out" scenario of 2080 is delayed; your children have a chance to live the life you have always wanted for them. In fact, if just one in ten homes used energy-efficient appliances, the effect on the environment would be like shutting down more than 30 of the dirtiest coal plants in the country.

Gas or Electric Stoves and Ovens

Gas or electric ovens are about equally expensive in terms of energy usage; the electricity comes from gas, or coal, burned at your local utility company. If your utility company is using alternative energy — such as solar, wind, or hydro — your electricity might cost more, but it is also more eco-friendly. Many utility companies will allow you to "buy in" to their alternative energy resources. The extra money goes to fund more alternative energy or research into alternative energy, and the extra $20 per month is pocket change when you consider the benefit to the environment. Programs that shut down or reduce your air conditioning electric use are of more benefit to the utility company managing "peak" loads than they are to the environment.

When it comes to cooking, many people prefer gas because it offers more controlled heat. If you choose gas, be sure to purchase an energy-efficient range hood to vent combustion by-products and cooking odors to the outside. Standing, or "always on," pilot lights in gas stoves more than double the appliance's annual energy consumption. If you have a choice, you should choose electronic ignition.

Self-cleaning ovens are more energy efficient because they have more insulation. However, if you use this feature more than once per month, you are defeating the added efficiency. Convection ovens, though more costly than conventional ovens, are the most energy efficient. The heated air is circulated continuously, so the heat distribution is more even, reducing cooking temperatures and times. Convection ovens save, on average, 30 percent of cooking costs over conventional ovens.

In terms of energy use, a conventional range, one designed with burners on top and an oven below, is the same as a wall oven and a cooktop; the only difference is aesthetic. To save energy, use a microwave to warm food, defrost food before cooking, or cook larger amounts at one time and freeze a portion. Invest in and use smaller cooking appliances; a toaster oven uses less than half the wattage of an electric oven and is fine for small cuts of meat. Soup made in a Crock-Pot set on low will cook perfectly and use less electricity than a coil on your stove because the Crock-Pot is insulated. Small appliances use about half the wattage of an electric oven because they heat a smaller area. When using, keep them out of drafts to reduce surface cooling of the appliance, which can rob it of its efficiency.

Refrigerators

A new refrigerator with an automatic-defrost feature and a top-mounted freezer uses half the energy of a refrigerator made about 1990. Side-by-side refrigerators use 10 to 30 percent more electricity than top-and-bottom refrigerators of comparable size, even if both carry the Energy Star label. This is because the government allows manufacturers of side-by-sides to assemble and sell them by different standards. Ice makers and see-through door models are less efficient. To compare energy across different models, look for the "kWh/year" information, either on the Energy Guide label on the machine or on the Energy Star Web site.

Refrigerators have a 15-year lifetime. If yours is nearing decrepitude, replace it. Size is an important energy consideration; models greater than 25 cubic feet use considerably more energy. Do not make the mistake of moving your old refrigerator to the basement and buying another; this will increase energy usage by a factor of almost two. Instead, buy a slightly larger refrigerator than your old one if it was inadequate and dispose of the old refrigerator properly. The store may pick up your old fridge for free or for a small fee. If this service is not offered then your disposal company will pick it up for a fee, or your city also might have a recycling program.

Microwaves

Microwaves are surprisingly energy efficient for two reasons. First, the energy heats up water molecules in the food, rather than all the molecules. Second, when used to heat or reheat small portions, the microwave does not have to be preheated, unlike an oven. Also, you can open the door of your microwave without losing heat. Most important, you can cook without heating up your house in the summer, reducing the need to turn up the air conditioning to compensate. Microwaves are now inexpensive; choose one with a turntable to improve cooking and energy use even more by preventing cold spots in food.

Dishwashers

The average dishwasher uses four kilowatt-hours of energy. Add in the heated dry cycle, and it climbs to ten kilowatt-hours. If your dishwasher was manufactured before 1994, replacing it with an Energy Star product can save you $30 per year in energy costs. These Energy Star-labeled dishwashers use 40 percent less energy than the federal minimum standard, use less water than older models, and use it more efficiently, resulting in cleaner dishes with less energy consumed.

Soil sensors can program your new compartmentalized dishwasher to run shorter cycles, reduce water temperature, and wash two different loads, such as glasses on top and pans on bottom, with equal effectiveness. You can select features such as light wash, a booster heater to relieve the load on your hot water tank, and air dry as opposed to heated drying.

Even if you cannot afford a new dishwasher, you can take steps to save energy. First, never run the dishwasher at less than full. Second, air dry your dishes by leaving the door open for an hour before you unload the dishwasher; on hot summer days, set a fan on the door. Third, if you must pre-wash, fill a sink full of water instead of leaving the hot water tap running.

Trash Compactors

A trash compactor uses an average of 1,400 watts, almost as much as a microwave. It lasts about 14 years. I had one and found it smelly and cumbersome. For about $10, you can buy a device that will flatten cans.

But, if you want to be truly eco-friendly, you should be sorting and recycling your cans, bottles, paper, and plastic, not flattening them. At $400, I rate a trash compactor expensive and pointless.

Clothes Washer

About 90 percent of the energy used to wash clothes is spent heating the water. The other 10 percent is used to agitate and spin the drum of the machine. Aside from using an inordinate amount of water, washing machines are relatively inexpensive, electrically speaking. If you always wash in cold water, the energy cost would be negligible.

You can maximize your old washing machine's eco-sense by setting the fill level to match the size of the load, washing in cold water, and letting the clothes sit in the washer with the lid open for an hour before drying. This allows excess moisture to evaporate naturally, which is especially beneficial in the winter, when your home is likely to be drier — leading to increased colds and respiratory complaints.

A new, front-loading Energy Star washing machine will save about 50 percent of the energy of your old machine, but 50 percent of 10 percent is only a 5 percent savings. What a new Energy Star washing machine will do is cut down your water usage and produce cleaner clothes with less agitation and soap, particularly if you choose a front-loading model.

Clothes Dryers

Clothes dryers have become a necessity in the modern world. Eighty percent of homes are equipped with them. In certain parts of the country, winters are too cold to permit effective line-drying. Many homeowners' associations forbid line-drying, and apartments and townhouses are not suited to it. In some areas, the local governing body has posted regulations prohibiting it, presumably for aesthetic reasons. That might soon change as electricity becomes more expensive and generating it becomes more difficult. In fact, a recent lawsuit against a homeowners' association that prohibits line drying might set a precedent for the recall of these kinds of statutes. In the meantime, we have clothes dryers.

After refrigerators, clothes dryers are the biggest electricity users. Gas dryers use half the amount of electricity and cost about 35 percent less to operate than electric dryers. You will not find this information on the Energy Guide label, because there is not one — dryers sold in the United States are not required to have them. According to the Department of Energy, there is too little variance between models to warrant an Energy Guide sticker.

Most dryers have a lifetime of 13 years. You can increase your old dryer's efficiency by locating it in a well-insulated room, keeping the lint screen and vent free of debris, using a non-flexible, metal vent of the shortest length possible, and taking your clothes out as they dry instead of waiting for the cycle to finish. You can even vent your dryer inside in the winter, allowing escaping heat to help heat your home. Cover the top of the vent with nylon to collect lint, and clean the nylon often, especially if you have a gas dryer.

You also can buy a new dryer. Even without energy ratings, the newer models offer a wealth of features designed to maximize drying while minimizing electricity or gas usage.

Television Sets

Television sets use a large amount of energy, even when turned off. The population of the United States is about 300 million. Most Americans have television sets, and most sets are on an average of five hours per day. Some households have more than one television. With about 260 million televisions running five hours per day, Americans are using an astounding 4.08 billion kilowatts, or 48 gigawatts, of electricity every month on television alone. That is half the production of a good-sized power plant and 1 percent of the nation's total power supply. These 48 gigawatts of production put more than 30 million tons of carbon dioxide into our air. Unless efficiency improves more rapidly than it has, television power consumption will double by 2025 due to more viewers watching on more television sets.

Ninety percent of the power is used when the television is on. Ten percent is used when it is supposedly off because manufacturers caved to consumer demand in the 1980s and installed standby power so the television screen would come on more quickly. Increased demand for larger screens and more advanced models, such as high-definition television, or HDTV, will further affect power consumption.

There is no Energy Star rating for televisions, but you can buy wisely — if you must buy. For a 27- to 35-inch screen, liquid crystal display (LCD) televisions are more efficient than cathode ray tube (CRT) televisions. For intermediate-sized screens about 40 inches, there is no clear efficiency standard because the technologies are still evolving. Projection televisions are most efficient when the screen size is 50 inches or greater.

For more information on this complex and evolving technology, go to **www.efficientproducts.org/tvs**.

Computers

Computers, like televisions, are "energy vampires," using power even when off. In 2006, Americans owned, or operated, about 200 million desktop computers and 100 million laptop computers, either at work, at home, or both. Fifty million new units are purchased every year. A desktop computer uses between 200 and 400 kilowatt-hours of electricity per year, a laptop about 100 kilowatt-hours. Many companies use desktop servers, which are simply a monitor and keyboard connected to a central mainframe. These use almost 2,000 kilowatt-hours per year — or as much per unit as you use to light your home. All together, computers and servers use 3 percent of all electricity produced in the country, and the Department of Energy expects use to continue rising into the next decade. In fact, computer use is one factor utility companies consistently take into account when planning future infrastructure, such as new power plants or new substations to distribute load. I used to work for a utility company, and the executives spent many hours agonizing about demand-side management — or how to keep the power on when everyone was at home running their appliances. The only advantage to the weekend, for a utility company, is that people are not in the office buildings with the lights on.

Ninety percent of computers now carry the Energy Star label, making it difficult to determine actual efficiency. Because of this — and because of the increasing sales of computers — the Energy Star program plans to revamp its computer specifications and will likely implement new criteria for calculating efficiency. However, until this happens, there are some simple things you can do to cut energy usage: buy the processing and memory speed you need and will use; you can buy backup dynamic random access memory (DRAM) or even a jump drive later if you need them; look for models that have the "80 plus" logo, a utility-funded program to cut energy use; look for energy-saving technologies such as AMD's "Cool n' Quiet" or Intel's "SpeedStep," which scale back power use when your computer is not being used; and remember, 33 percent of the energy is going to your monitor, so turn the screen off or put the computer on "sleep" every time you leave your desk.

Air Conditioners

The average central air unit uses 3,900 kilowatt hours (and emits 7,000 pounds of carbon dioxide) in a normal cooling season in the Midwest, as compared to 1,000 kWh for a standard-sized window air conditioner. The window air conditioner only cools one room, so most houses require about three (one for the main area and one for each bedroom), bringing the total kWh to 3,000 for an inefficiently cooled home. If you live in the cool mountains of central California, you probably will never need air conditioning. If you live in Florida, you will. In the Midwest, where summers are sultry, you will need air conditioning during July and August and possibly more often. You could substitute a dehumidifier and fans, but the electricity savings is minimal compared to the decrease in comfort.

If you have an older central air conditioner, you might want to replace the compressor with a new, high-efficiency unit, for a net electrical savings of about 40 percent. A new compressor will cost you about $900, not including installation.

Buying the right size and having it professionally installed are the key components in air conditioning efficiency. If your house is 2,200 square feet, you need a five-ton compressor, as one ton is needed for every 500 square feet. A one-ton compressor is equivalent to 12,000 Btus. Window units are measured in actual Btus. If you buy a unit that is too large, you will have excess humidity; too small, and it will not cool your living space.

Energy efficiency is rated in Seasonal Energy Efficiency Ratios (SEERs). Older units might have a rating of six or less. Today, the minimum permitted rating is 13 for any unit made after January 2006. Look, as always, for the Energy Star label, and buy a model with ratings above 13 if you can afford it. The extra $500 to $1,000 you spend will pay for itself over the life of the unit.

Also look for such features as a thermal expansion valve, a variable-speed air handler, a fan-only switch, a filter-check light, and an automatic-delay fan switch. If possible, check the decibel rating of the unit in operation; you might not care, but your neighbors might. A new unit will have a warranty, but you need to read it carefully to see what is and is not covered under the terms of the sale and installation.

Freezers

Chest freezers are more energy efficient, and thus cheaper to run, than upright freezers. Unfortunately, they also are harder to get food from and harder to defrost. Even then, the differences between styles are not substantial; the real difference involves size. A small upright is much more energy efficient than a giant chest freezer. A 17-cubic-foot upright uses about 200 kilowatt-hours per month. Your utility company charges you more for every kilowatt-hour you use, but a good rule of thumb is $.10 per kilowatt-hour. A chest freezer reduces that usage by about 20 kilowatt-hours, or $2 per month. Adding the frost-free feature will add about 40 kilowatt-hours, or $4 per month, to your operating cost and should be avoided if possible.

Keeping your freezer in the garage will not save energy because it has to run during the summer, too. Frigidaire does have a line of garage-friendly appliances designed to withstand extremes of cold and heat, but these feature internal heating and cooling systems, which will only add to your electric bill and your carbon footprint. Experts recommend keeping freezers in the basement, where temperatures are moderate year-round. If you do not have a basement, choose an insulated porch or some area in your home not subject to extreme temperature fluctuations — every time the compressor has to kick in, your electric dial spins more.

Freezers run most efficiently when filled, so open the door once a day and select all the frozen food you will need for 24 hours. Do not buy a freezer larger than you need, or you will be cooling empty space. Select a model with an adjustable temperature control, and monitor the temperature regularly of the room the freezer is in. Choose a model with an automatic lock if you have children, as children have been known to hide inside freezers. Some models have a power light telling you whether the freezer is plugged in and operating; this is valuable if you have frequent outages that might shut down electrical circuits. If your power does go out, the food will keep for 48 to 72 hours, as long as you do not open the door.

The Maytag MQU1654BEW, an upright freezer, and the Whirlpool EH151FXR, a chest-type freezer, have earned the Energy Star seal of approval.

Prices range from $400 to $700. Buying in bulk and freezing food is an excellent way for a middle-class family to cut the food budget, but do the math first. Unless the cost of the freezer, averaged over 14 years, and the cost to run it are less than the amount you would spend to buy the food fresh, you are not saving money, and you are adding to your carbon footprint.

Hot Tubs

Hot tubs are an extravagance few of us can afford. For those who can — or who need the therapeutic effects of soaking in hot water — hot tubs are an expensive proposition. The Internal Revenue Service (IRS) might or might not consider it a tax-deductible expense on your annual tax return, depending on its size, location, and your proof of need. To be counted as an expense on your yearly tax return, the cost of the tub must exceed 7.5 percent of your adjusted gross return, and you cannot turn around and deduct it as a home improvement on your capital gains return when you sell. Your insurance company will have similar stipulations, and you might not recover the entire cost. Medicare will not cover it because you could obtain the hot-water therapy at a gym or licensed medical facility.

Hot tubs cost, roughly, about $80 per month for a four-person, outdoor, insulated model. The determining factor is your climate; if you live in Minnesota, you might spend a great deal more in winter. If you move the hot tub inside, you reduce your costs by $60 and your carbon footprint by 80 pounds of carbon dioxide per month. Newer electric models are nearly twice as efficient as older models; there is, however, no Energy Star rating on hot tubs, spas, or whirlpools. Indoors or out, you will need a dedicated electrical outlet and a convenient way to get water into the hot tub — this does not include a pail. You also will have to spend a certain amount of time and money maintaining, and possibly winterizing, your hot tub. The cost is minimal, but the time and effort required are immense. If you need a hot tub for therapy, you probably do not have the energy to maintain it and would be better off joining a gym, which would further reduce your carbon footprint, especially if you use public transportation to get there. For more information, go to the IRS Web site at **irs.gov**, to **arthritis.about.com/od/taxes/f/poolsspas.htm** or **www.hurt911.org/therapeutic/hot-tub-insurance-tax-deduction.html**.

Pools

An outdoor, residential, one-family pool about 16 feet wide and 28 feet long uses enough energy in one summer to power your home for three months. A pool uses 1,500 kilowatt-hours during one month of operation and adds $150 to your electric bill and a whopping 3,200 pounds of carbon dioxide to the atmosphere.

However, if you must have a pool, pay particular attention to the pump and heater. If you reduce filtering time to four or five hours per day, you can save almost 40 percent of electrical costs. Consider investing in a timer to regulate this, and set the pump to run several hours in the morning and the balance of time in the evening — or vary the schedule according to your family's needs.

You can heat your pool using the sun's rays. A simple solar blanket will run from $30 to $300, depending on the thickness of the plastic and whether it is coated with aluminum. These covers have to be removed to use the pool, and small children may be caught under them. If you live in a fair-sized city, you will have to fence or otherwise secure the pool for safety, but this might not protect your own children. Drowning is the second leading cause of home-related child deaths.

Solar heating systems are safer but cost between $2,000 and $4,000 to install. This cost should pay for itself in two to seven years, depending on your utility's rate structure and the climate where you live. For more information, go to **www.consumerenergycenter.org/home/outside/pools_spas.html**. For more on solar power and lighting, read the Case Study on page 24.

Heated Waterbeds

Heated waterbeds are luxuries for everyone but the ill and elderly, and they can cost between $15 and $30 per month to operate, depending on your climate, how well you heat your house, and their temperature ranges. They use about 200 kilowatt-hours per month and contribute five pounds of carbon dioxide to your footprint. You can decrease that footprint by setting the temperature lower or adding extra blankets to the bed when you are not in it to conserve heat and energy. Of the three choices — hot tub, pool, or heated waterbed — I would recommend the waterbed. It is the only location where you spend a full eight hours.

Electricity: "Energy Vampires"

"Energy vampires," those sneaky appliances that lie in wait, sucking electricity even when you think they are sleeping, are costing American consumers about $3.5 billion per year.

It is not an expression you want to use around young children, who will plead for a night light, driving your electric consumption even higher, but energy vampires exist; your set-top cable box uses almost as much electricity off as on, as does your answering machine and your battery charger. Microwaves, computer printers and scanners, video cards, and televisions all quietly suck power even when they are not working for you and slyly consume 10 percent of all electricity produced, costing the average American about $80 per year.

Called "phantom" energy, this silent consumption of electricity requires 65 billion kilowatt-hours of electricity every year and contributes 87 billion pounds of carbon dioxide to our already troubled atmosphere. Your personal share of this is four pounds of sulfur dioxide, seven pounds of nitrogen oxide, and a whopping 2,400 pounds of carbon dioxide each year. A tree will absorb roughly 50 pounds of carbon dioxide per year. At the rate you are using phantom electricity, you will need to plant about 40 trees per year to offset the energy vampires in your home.

The biggest culprits are adaptors, such as those on cordless phones, digital cameras, and hand-held power tools. Most are using power when they are plugged into an outlet, even when they are fully charged and ready to use. The second-largest group of energy vampires is electronic devices that operate in standby mode, such as computer monitors. You can identify many of them by their LCD display, which shows a light or the time.

Designers and manufacturers incorporated these features, such as auto-on and standby, for the convenience of the American consumer. We are now at a crossroads where convenience and sustainability collide. Here is what you can do to help.

- **Unplug** — If you are not using it, do not let it use energy.
- **Read** — The label will tell you how much standby power is required. If you choose Energy Star products, they will use less standby power. Alternatively, visit the Department of Energy's Web site on product specifications at **www.energystar.gov/index.cfm?c=product_specs.pt_product_specs**.
- **Simplify** — If it has many fancy options, it is going to use more energy, and you will have to expend more energy working to pay for it, plus planting yet another tree to clear your carbon footprint.
- **Just say no** — Thinking of buying several televisions or computers for the kids' bedrooms? Instead of two or three televisions, consider setting up a central area where all the junior members of your family can share these amenities. You could put the extra money in a college fund, and sharing will teach your youngsters the meaning of teamwork.

To learn more about energy vampires, go to **www.ucsusa.org/publications/greentips/energy-vampires.html**.

Water: A Finite Resource

Most people think of water as an infinite resource, especially those along the Atlantic Seaboard, the Gulf of Mexico, the Upper Midwest, and the Pacific Coast north of California. In these areas, the rain is predictable, lakes and rivers continue to exist and flow, and all one has to do is turn on a faucet to see the proof of this seemingly infinite resource.

In fact, as the inhabitants of Arizona and other arid regions now realize, water is a finite resource. Ninety-nine percent of the earth's water is tied up in oceans, in water too saline to drink or use on crops or in gardens. Seawater contains 130 grams of salt per gallon. To use it in agriculture, the salt content must be reduced by 125 grams per gallon. The salt content must be further reduced, to less than two grams per gallon, before it is considered safe for human consumption.

The average American uses between 80 and 100 gallons of water per day. The country as a whole uses about 323 billion gallons per day. Most of this is surface water, but the other 85 billion gallons comes from the ground. Some people argue that groundwater should not even enter the equation, but a realist accepts that groundwater must have, at some point, come from the surface and is therefore another finite, rather than infinite, resource.

According to an article on LiveScience (**www.livescience.com**), desalination is not economically viable at this point. Evaporative methods use tremendous amounts of energy. Membrane technology, though rapidly improving, still requires an inordinate expenditure of energy in relation to the product; it takes about 14 kilowatt-hours of energy to produce 1,000 gallons of potable, or drinkable, water. Whether using evaporative technology or membrane technology, the United States would have to add

100 more power plants, each with a billion watts of capacity, to provide half the water used in this country over an average year.

With water use growing twice as fast as the population, scientists cannot even estimate what will constitute an average year in, say, 2010. In fact, they cannot even say what constitutes a drought. Phoenix, for example, gets about 8 inches of rain a year. In 2006, it was a meager 5.25 inches. According to some predictions, the American Southwest is in for 90 years of drought as a result of global warming.

Scientists can, however, say with some certainty that the rising cost of — and increasing scarcity of — water will make desalination economically feasible in the next ten years. This is a negatively charged prediction, sort of like telling your balding friend he would look better if he shaved all his hair off.

Until desalination becomes economically viable, we are living on borrowed time — and borrowed water from rivers and lakes whose depletion directly affects the ecological balance of our planet. In ten years, we will face a similar predicament, taking water from a seemingly infinite ocean whose ecological balance is equally perilous. We cannot continue to spend these resources with no regard for our children's future or the future of the planet.

To determine the size of your water footprint, go to: **www.waterfootprint.org/index.php?page=cal/waterfootprintcalculator_indv**.

Water: Using it Wisely

According to the EPA the average American wastes up to 30 gallons of water every day. It might be as small a problem as a leaky faucet or as large as a sprinkler that runs in the rain. Either way, the EPA predicts that 36 states may face water shortages as early as 2013.

Every day, this country uses 323 billion gallons of water. The largest users are power plants, followed by agriculture, manufacturing, and other applications. Household use accounts for 11 percent, or about 26 billion gallons. If everyone followed the recommendations below, we could cut that usage by at least 30 percent.

- Install low-flow aerators in all water faucets, and do not run the water while brushing your teeth. Do not run water to get a glass of water; if you want it colder, use ice. To learn more tips, and get advice on low-flow faucet aerators, go to **eartheasy.com/live_lowflow_aerators.htm**.

- A five-minute shower uses between ten and 25 gallons of water; a tub bath takes 70. Installing low-flow shower heads saves 50 percent more water and the electricity to heat that water.

- A leaking toilet wastes 200 gallons of water per day. To test your toilet for leakage, put green food coloring in the top tank. If the toilet water turns green, you have a leak. A simple toilet repair kit costs about $25. Water costs about $2 for 1,000 gallons, but the cost is rising daily. In two months, you will have paid for the kit and cut water costs. Just be sure to shut off the water at the toilet valve, located near the floor, before you start.

- Older toilets use between three and five times as much water per flush as newer, high-efficiency models. You can install a water displacement device in your old toilet; usually a plastic container or plastic bag that displaces water normally flushed away. To find out how, go to **livinggreen.ifas.ufl.edu/water/water_conservation.html#devices**.

  Check to be sure that the displacement device does not interfere with the toilet's operation.

- Do not water your lawn when it is raining.

- Do not run water to wash your dishes. Instead, fill the sink. If you use a dishwasher, do not run it until the baskets are full.

- Set your clothes washer's fill level to match the size load you are washing. Do not wash ten items in a full tub. Wash dark colors in cold water to keep them looking new longer. Use cold water to rinse all loads. Turn down your water heater; your white sheets will still be as white, and you can afford to change them more often.

- Leaky faucets can waste up to 3,000 gallons of water per year. Fix them. For instructions, either in text or video format, go to **www.ehow.com/how_15854_fix-leaky-faucet.html**.

If your faucet is not leaky, check your toilet, swamp cooler, water softener, or ice machine. To check for a leak, read your water meter at night, after everyone is asleep, and again in the morning. Except for an occasional nocturnal flush, the reading should not change. If it does, turn off the water at the main intake where water enters your home and wait an hour. If the water meter still registers usage, you have a leak. Your local water company will come out to read and check your meter, but you will need a licensed plumber to repair leaking inflow pipes.

Toilet Eco-sense

Americans use vast quantities of water inside our dwellings — more than twice as much as Europeans and four times as much as the average Chinese family. In the United States, the average family uses about 280 gallons of water per day, most of it in the bathroom. Toilets use more than 25 percent of that water.

Older toilets use between three and five times as much water per flush as newer, high-efficiency models. These newer standards are federally mandated. To read more about federally mandated standards and their origin, read the study at: **www.cuwcc.org/Uploads/Tech_Docs/PlumbingStdsPaper_02-05-17.pdf.**

If your toilet is old, it may be time for a replacement anyway. Most water-efficient, or low-flow, toilets are priced between $150 and $500, the latter featuring pressure-assisted or vacuum-flush models. For more information on environmentally friendly toilets read the Case Study below, which describes a simple modification for an older toilet that can reduce water usage.

CASE STUDY: AGAINST THE FLOW

How Homeowners can save Thousands of Gallons of Water each Year

By John Kirkland,
Media Representative, Athena Company

There is a home improvement dilemma facing millions of homeowners. "When do I change out my large tank toilet for a new 1.6 gallon toilet?" Homeowners are consistently faced with increased water and sewage costs, and many see the advantages of the new low-flow toilets, but at hundreds of dollars to switch, many families would rather wait.

Now a new product is hitting the national market that makes the switch quick and easy — The Controllable Flush™. The Controllable Flush was used and tested extensively during the '90s, and the Athena Company is introducing a newly redesigned Controllable Flush.

The Controllable Flush is a simple addition to the large tank toilets that hold 3.5, 5, or 7 gallons of water. The handle mechanism allows 1.5 gallons of water to be dispensed when pressed down and the entire tank emptied when pushed up. In terms of home improvement ease, The Controllable Flush could not be any easier. In less than five minutes, the average family will be saving money and water almost immediately.

Here is the amount of water (in gallons) the average family will save using the Controllable Flush:

Persons per household	Year 1	Year 2	Year 3	Year 4	Year 5
2	15,330	30,660	45,990	61,320	76,660
3	22,995	45,990	68,985	91,980	114,975
4	30,660	61,321	91,980	122,640	153,300
5	38,352	76,650	114,975	153,300	191,625

CASE STUDY: AGAINST THE FLOW

The chart shows significant savings in water, which translates into lower water and sewer bills. Using the Controllable Flush is easy — pull up on the lever for a full flush, push down on the lever for 1.5 gallon flush. The homeowner has control of the water consumption and can decide water usage, as opposed to being left with no option per flush.

"Home improvement options that save money should not be a hassle," said Controllable Flush developer Steve Lord. "The Controllable Flush is one of those inexpensive products where you can do your part for the environment, save money and water, and it's easy to use and install. That's why we've sold thousands over the past few years and why we are rolling out a national push to help other communities," noted Lord.

In times where water has become a valuable commodity and an increasingly expensive luxury, it makes sense to take steps to limit waste. Currently the United States uses almost 2 trillion gallons of water per year. It's time to start using water smartly.

To date, the Controllable Flush has been endorsed by communities throughout the West, including: City of Sacramento, California; City of Las Vegas, Nevada; city of Jefferson, Oregon; and the State of Nevada.

Purchasing The Controllable Flush is simple. Go to: **www.controllableflush.com** or call 1-888-426-7383. To see the Controllable Flush in action, go to **www.athenacfc.com/howitworks.html**

About Athena: Athena is a U.S.-owned and -operated company with offices and manufacturing facilities in Portland, Oregon.

The Controllable Flush Handle is patent protected, manufactured under strict ISO 9002 certification and is truly "Made in the USA."

The Controllable Flush was invented in 1986 by Bob Moore, a Portland boat owner who was looking for a way to save water in his house and on his old cabin cruiser. After going through several prototypes, he submitted the Controllable Flush to Stevens Institute of Technology for testing in 1992. The Institute determined that the Controllable Flush was one of the best retrofit water-saving devices in the United States. Athena president Steve Lord bought the patent in 1996 and today manufactures the Controllable Flush.

We believe our products offer an easy and effective way of helping to conserve one of nature's most valuable resources: water.

Case study courtesy Steve Lord, Athena Company, Oregon

Lawn Eco-sense

If you have to water your lawn, buy a droplet sprinkler instead of a mist sprinkler. Mist sprinklers deliver a fine spray that, on a hot day, evaporates almost before it reaches the surface. This mist also conducts sunlight more strongly, burning your grass or ornamentals when the sun is highest. Droplets evaporate more slowly and do not conduct as much light. If you doubt this, consider the refraction potential of 1,000 pieces of glass or crystal as opposed to ten.

There are several types of lawn sprinklers: stationary, which sit on a spike; rotary, which throw water from circulating arms; oscillating, which involve a long arm that moves back and forth; and pulsating, or impulse sprinklers, which deliver individual jets of water in rapid succession over an arc. Oscillating sprinklers are most common. Whichever you choose, make sure it delivers droplets and not mist, which, in addition to its rapid evaporation and refraction tendencies, is easily deflected by a modest breeze.

You might want to invest in a traveling sprinkler, which "walks" your lawn on a guide, delivering water to different areas without you having to move the hose. If you have a new home, or resent the time it takes to water your lawn, you might consider an in-ground sprinkler system. The problem with most sprinkler systems is that they cause runoff, waste water by over-watering in unwanted areas, require much digging and installation of various heads, cause water-pressure problems inside the home, get plugged or damaged by lawn mowing or freezing, and are poorly suited to watering plantings taller than six inches. They also require a large amount of line and parts and usually demand professional installation, which can be costly.

Whatever you choose, use a timer — either a simple oven timer to remind you that it's on or the kind that attaches to your outdoor water faucet. Aubuchon Hardware online has a variety of timers, from standard to analog, for manual and in-ground systems, from about $15 to as much as $75. I have purchased from them and found them to be fast, friendly, and fair.

Use a moisture sensor, either the first time you water to determine precisely how much water your sprinkler is delivering, or install sensors permanently. Aubuchon sells an in-ground water sensor called Wormy, but it has to be purchased in bulk if bought online; Aubuchon retail stores are in New England. However, at about $65 for 24 devices, they are still cheaper than soil-moisture sensors that come with a spike that, inserted into the soil, gives you a soil-moisture reading. Be sure you have bought the appropriate sensor for your system; automatic systems use a type that detects both soil moisture and precipitation to prevent watering when it is raining, but they will not work with manual sprinklers.

Dishwasher Eco-sense

If you have an older dishwasher, fill it completely before running it. Do not use the pre-rinse, rinse-and-hold, or heated drying features. Use the air-drying option instead. If your dishwasher does not have this option, leave the door open, and the dishes will dry on their own in a few hours. If you have not already turned your hot-water heater down a notch, do so now. Put a quarter-cup of baking soda in a mesh basket in the lower tray of the dishwasher. Your dishes will be just as clean as if you had used hot water and will be less likely to spot. You also might want to consider installing a flow-reduction device on the water line, though I have found these to be prone to failure.

Contrary to popular belief, dishwashers are energy efficient. They use 37 percent less water than hand-washing dishes. This is because most people run the hot water to rinse dishes. If people would fill one side of the sink with rinse water, hand-washing would become more energy efficient.

Older dishwashers require water hot enough to dissolve grease, or 140 degrees. Newer dishwashers have a booster heater, which allows you to set your hot-water heater closer to 120 degrees — hot enough to clean people, if not dishes.

Like washing machines, most of the energy your dishwasher uses goes to heat water. By saving water with a new, energy-efficient dishwasher, you are saving not just the cost of energy and water, but also the cost to treat it in a municipal facility, transport it to your home, meter and bill for it, and clean it after use. Up to 50 percent of a typical city's energy bill goes to supplying water and cleaning it after use.

If your dishwasher is close to 11 years old, its normal lifetime, consider investing in a new, water-and-energy efficient model. Energy Star-approved dishwashers cut water usage by 30 percent and electricity costs by $30 per year.

Clothes Washing Eco-sense

Washing machines, like toilets, are notorious water hogs. An average family uses roughly 16,000 gallons of water each year to clean clothes. This represents $32 per year just for water. Add in the cost of electricity to run the washing machine and the cost of heating water and you have accounted for a significant portion of your energy and water bills.

Top-loading machines are less efficient than front-loaders. A typical front-loading machine uses about 50 percent less water, 40 percent less energy, and half as much detergent. In addition, front-loaders spin faster, so the clothes contain less water when put in the clothes dryer and are quicker to dry. Unfortunately, manufacturers in this country focused on the top-loading machine until the middle of the last decade, so consumers had little choice when selecting a new washing machine.

That trend has changed. New machines with the Energy Star label feature many front-loading washers, often sold as a set with an energy-efficient dryer in avant-garde colors such as burgundy, for about $800 each. The price is high, but the savings are ongoing for the life of the machines. Energy Star washing machines use only 18 to 25 gallons of water per load, compared to 40 gallons per load for older models, and save about 7,000 gallons of water per year.

If you are building a new home, consider asking your plumber to rig the pipes so you can water your lawn and flowers with "gray water," which is taken from washing clothes and dishes, taking showers, and bathing. It is mostly a matter of redirecting the outflow drains to a holding tank instead of a septic system or the sewer main. Gray water use can save from 25 to 40 percent of your water by recycling it, and it is ideal for lawns, flowers, trees, and shrubs but should not be used on anything that will later be eaten. In other words, you can use gray water to water your apple trees, because the tree "filters" the water, but you cannot use it on vegetables. It should never be allowed to form runoff and should not come from soiled diapers or feces — this is called black water — or from meat preparation or anyone with an infectious disease. You might have to change some of your buying habits, including switching to eco-friendly detergents and brighteners, soaps, and cleaners, but this is part of being green. To learn more about gray-water usage, go to **www.wvu.edu/~agexten/hortcult/homegard/graywate.htm** or **interests.caes.uga.edu/drought/articles/gwlands.htm**.

Gardening Eco-Sense

No one knows who wrote, "She who plants a seed beneath the sod and waits to see a plant believes in God," but most gardeners agree something miraculous occurs in a garden — if not in the actual vegetation, then certainly in the spirit. Gardening is one of mankind's most hopeful enterprises and implies by its very act that life will be consistently renewed. Given that, and a gardener's appreciation of the natural world, it is hard to imagine how some gardeners go wrong, but they do.

It is hard not to love grass, that green carpet on which all our hopes are pinned, our aspirations revealed, our spirits renewed. Why, then, do we — nature and grass lovers — persist in over-watering our lawns, using fertilizers and pesticides, and using a gas-fired lawn mower that produces more pollution than driving a car?

Watering Your Lawn or Garden

Drought in Georgia, Florida running out of water, the Great Lakes shrinking, the Sierra snowpack melting faster every year, the Colorado River drying up — all these symptoms are part of a larger picture. We are reaching the point, as a nation, when we may no longer be able to provide one of life's essential elements. The government says that at least 36 states will face water shortages in the next five years as a result of drought, expanding population, and, most important, waste. The price — for conservation, recycling, desalination, and stricter controls — will be staggering; $300 billion over the next 30 years just for infrastructure. Cheap water will become a thing of the past.

As a nation, we use about 280 gallons per household every day. This is three times as much as China, more than twice as much as India, and twice as much as any of the developed countries in Europe. Agricultural water accounts for 40 percent of the United States' water use. Generating electrical power, whether via generators, nuclear, or hydroelectric, accounts for another 39 percent. Industry uses 5 to 6 percent. Raising livestock or mining uses only 1 to 2 percent. Public water, the water you drink and bathe in, accounts for 13 percent. To provide more water to homes, we will have to take from the large users, such as agriculture or power suppliers. The first choice, from a local or federal government standpoint, is always agriculture. There are two obstacles to this: first, many agricultural users have water contracts with federal, state, or local governments that cannot easily be broken, and second, taking water from agriculture reduces our food supply.

In a report commissioned by the United States Bureau of Reclamation, the council noted, "The combination of limited water supplies, rapidly increasing populations, warmer regional temperatures, and the specter of recurrent drought point to a future in which the potential for conflict among existing and prospective new water users will prove endemic."

Is it suggesting water wars? Yes. The sad part is, these could be prevented if more homeowners used common sense. According to the EPA, one-third of all residential water use goes toward keeping lawns and gardens green, and much of this water is lost to runoff, evaporation, over-watering, and inefficient landscape design.

If you live in an area of relatively abundant water, you still need to make some choices. These choices should include native plants, which are adapted to regional temperature and rainfall patterns. You should group your plants by their water needs; in the Midwest, coneflowers and coreopsis thrive without much extra water, but roses do not. If you have treated your flower beds with compost, water will be naturally retained. In clay soils, which hold too much water, compost helps excess water drain into underground aquifers instead of rotting the roots of plants or washing into storm drains. Mulch also will help retain water and keep the soil cool and damp, as well as prevent the growth of weeds. As mulch rots, it becomes compost, further benefiting your plants, but never spread it more than a few inches deep, as this deprives the soil of needed oxygen.

When you do need to water, use soaker hoses or drip irrigation placed on the soil and under a layer of mulch. These deliver water to the roots of plants, preventing both evaporation and runoff. On a hot summer day, evaporation can account for 30 percent of water loss and raise the ambient humidity in your yard several degrees, leading to mold, mildew, and pest infestations.

If you have grass, consider converting unused areas of your lawn to native grasses and flowers. Plant drought-resistant varieties of grass, and let them grow as high as local law permits; the blades will shade the soil, further preventing evaporation. For awkward, hard-to-reach, and difficult-to-water areas, put down decorative gravel interspersed with statues or small oases of flowers planted in a wooden tub sunk in the ground. The effect is aesthetically pleasing, and you can carry a pail of water instead of dragging a hose around. For more information on landscaping with native plants, go to **plants.usda.gov**.

If you live in the arid Southwest and want to reduce your environmental footprint, do not plant grass. If you need grass, move to the Midwest or the Northeast where water is more plentiful. All choices represent a trade-off. Sustainability is a trade-off between inhabiting a planet that remains viable for generations and making plastic from petroleum. For Southwestern gardeners, xeriscaping is an excellent method of cultivating the earth without impacting already perilous water supplies.

Last, consider collecting water via the rain barrel method. Jennifer Carlson of Haven Illustrated sent me her observations on the benefits of rain-barrel collecting, and these are provided next.

CASE STUDY: JENNIFER CARLSON

Here is the setup that we have for our 625 gallon cisterns. Each is buried in a bed of sand 2 1/2 feet below grade (for earthquake stability). Each has a 1-inch overflow connector to the next cistern, then the overflow from the winter excess flows to the gravel driveway, which is at the lowest point on the property. We have spigots attached to each one for watering cans and hoses.

There are inexpensive sump pumps that are built with enough power for irrigation with a 3/4-inch hose connector to a lawn sprinkler or soaker hose. We are debating on whether we want an electric or solar unit. So far, we have never used all of the water we gather.

There are three cisterns. We harvest from half — or one side — of our 1,000 square-foot roof. The downspout runs along the fence line into the first cistern. We did not have the cisterns installed when *Seattle Homes and Lifestyles* magazine wrote an article about us, but it featured other ecological design elements in our garden. We have been featured in several articles:

- *Organic Gardening* magazine, "Beak Experience: Chicks in your Garden," June/July 2005
- *Seattle Homes and Lifestyles* magazine, "Teaching by Example," October 2005
- *Natural Home and Garden* magazine, "City Chicks," May/June 2006

CASE STUDY: JENNIFER CARLSON

- "Chicken Coops," by Judy Pangman, Storey publishing, ©2006

Courtesy of Jennifer Carlson, Haven Illustrated, LLC

Xeriscaping

Xeriscapes are not a monotonous landscape of rock and stone. They can be unique expressions of the homeowner and the local climate and can cut water use more than 50 percent compared to normal gardening. They also can reduce the time you spend maintaining your yard, eliminate or reduce the need for pesticides and fertilizers, create a safe zone around your property if you live in areas prone to fire, reduce air pollution by removing the need for lawn mowers and other small-engine yard maintenance tools, and promote local wildlife — everything from butterflies to birds, frogs, lizards, and other habitat-threatened species.

Before you attempt remodeling your yard as a xeriscape, you should draw a diagram of the space and spend several days or weeks noting which areas receive the most sun; which areas lend to runoff; which areas have good soil; which areas are already allotted for walkways, drives, and patios; and which areas nearest the house would provide natural cooling, via shade, if trees and shrubs were planted. Xeriscaping is simply feng shui for yards. It allows nature and aesthetics to plan your garden, instead of relying on your own, sometimes limited, understanding.

Mark your diagram with the points of the compass. Areas that will remain turf should be designed for easy mowing, with curved rather than straight edges. Include all existing trees and shrubs that you plan to keep, as well as all existing or planned paths, patios, and structures.

Amend the soil where necessary. Cacti and other succulents will tolerate sandy soil, so if you have much sand and do not want to put in much work, stick with succulents. This choice does not limit you as much as you might think; there are hundreds of varieties of succulents in all sizes, shapes, and bloom colors. Many grow wild; others are cultivars of wild plants and perform beautifully in the hot, arid climate of Phoenix and other Southwest cities.

Buy and install your plants. If you have chosen some species that will need more water than local rainfall provides, put down a soaker hose. Use an organic mulch around all but cacti. Use decorative rock or gravel in other areas. When you water, water deeply. This encourages roots to grow into the lower levels of soil, where water remains long after it has evaporated from the surface.

There are a few more essential elements in good xeriscaping, such as fertilizing, composting, and pesticides, and these are covered here too.

Rain Gardens

Polluted runoff results when an area that receives much rain or snow is urbanized, resulting in roadways, sidewalks, driveways, and other hard surfaces that contain man-made pollutants such as fertilizer, pesticides, oil, and car exhaust. Water picks up these pollutants and, unable to percolate below these hard surfaces, runs into storm sewers and from there, into rivers, lakes, and streams, taking all the pollution with it.

Rain gardens, by percolating water through earth, are landscapes that improve local water quality and recharge groundwater supply while providing habitat for birds, butterflies, and other natural denizens of wild places. Rain gardens benefit the ecology and the cities in which they exist by providing attractive places for people to visit, inside what would otherwise be a steel-and-concrete prison. Sociologists have long known that creating "urban oases," such as parks and other natural features, relieve the stresses of urban life.

A rain garden, whether in your yard or in your community, can provide that relief. Rain gardens are not difficult to build. The typical rain garden requires a depression in the landscape to collect rain and snowmelt. This can be as large or as small as you want, but keep it at least 10 feet from your house to prevent seepage that can ruin your basement or foundation. Also, resist the temptation to locate it in a place where water already "puddles," as this indicates an area where water migration to deeper layers is already slow. Put the garden in sun, either full or partial, but not under a tree, as it will damage the tree's roots. Locate it in an area where water is provided by downspouts, runoff from walks or drives, or other natural water collection devices. If you live in an area of heavy rain or deep snow, provide a drainage ditch or culvert to another rain garden or to the street.

When you have created your depression, removing excess soil and loosening the remaining soil or adding mulch, you can edge it with sod, shrubs, or stone to prevent it from looking like an accidental puddle, or you can leave it alone. In a rain garden, water is designed to percolate to lower layers of soil within hours. Choose plants that do not mind "wet feet" at intervals. Always choose native plants, a list of which can be found on your regional Department of Natural Resources Web site.

Contrary to popular belief, rain gardens do not attract mosquitoes. Mosquito eggs survive only in water that stands for a number of days; most rain gardens dry in a matter of hours. Water-loving native plants can survive in dry conditions as well, so resist watering them during dry spells, or they will get "leggy," or too tall, and droop in a unattractive fashion.

In addition to roadways and other hard surfaces, most urban and residential areas are primarily grass. Grass, or turf, does not percolate water well and requires regular application of fertilizer and water to continue looking attractive. A rain garden, which is in essence a "mini-wetland," mediates these problems, provides habitat, and soothes urban anxiety.

There are a number of wildflowers and shrubs that do well across most of the United States in rain gardens, and they include

- Black chokeberry
- Blueflag iris
- Boneset
- Canada anemone
- Cardinal flower
- Giant hyssop
- Goldenrod
- Great blue lobelia
- High bush cranberry
- Low bush honeysuckle
- Marsh marigold
- Marsh milkweed
- Mountain mint
- New England aster
- Prairie phlox
- Pussy willow
- Red osier dogwood
- Softstem bulrush
- Tall meadow rue
- Virginia bluebells

Fertilizers

There are two types of fertilizers: inorganic and organic. Inorganic fertilizers are granules or liquids made from nitrogen, phosphorus, and potassium. Organic fertilizers include compost, cow or sheep manure, moss, bark, or other natural materials, and contain nitrogen, phosphorus, and potassium but at lower ratios than inorganic fertilizers. The advantage of organic fertilizers is that they work slowly, in harmony with nature, by encouraging microorganisms that benefit plants.

Inorganic phosphates, found in fertilizer, are a major source of the pollution found in lakes and streams. Called eutrophication, this enrichment of surface waters with phosphorus and nitrogen results from runoff and soil erosion, and it causes an increase in undesirable aquatic plants, such as algae and milfoil. The algae bloom, creating toxins that cause a bad odor and taste in the water, and then die, consuming much-needed oxygen from the water and impacting the fish population. The plants alter the habitat, and "good" fish such as walleye die off, while "bad" fish such as perch continue to thrive. Fishermen abandon the lakes because the weeds make it too hard to operate an outboard motor and the fish caught are inedible. The DNR also abandons the lake because budget cuts mandate focusing on a few notable lakes. In time, the lake dies. Herons, egrets, fish gulls, and eagles move elsewhere. Something is lost that can never be regained.

In the Gulf of Mexico, off the coasts of Louisiana and Texas, a dead zone appears every spring. In 2007, it was 7,900 square miles, the third largest since recording began in 1985. The cause is fertilizer runoff from as far away as the Corn Belt, on the upper Mississippi. In this zone, oxygen levels drop too low to support aquatic life, threatening the half-billion-dollar Gulf fishing industry. Increased production for the ethanol industry, which makes corn into fuel, will only intensify this problem. Donald Scavia, a University of Michigan scientist who led the first federal, integrated assessment of the Gulf dead zone in 2000, admits there is no way to control its spread unless the agriculture industry bans such widespread and indiscriminate use of inorganic fertilizers.

If you fertilize your lawn, then cut it; you are contributing to the Gulf dead zone. The fertilizer sticks to the blades of grass and leaves as excess, and the first good rain washes this excess into the sewer. If you must mow, and most city ordinances indicate you must, use a mulching mower. Better yet, start a compost pile. The excess fertilizer will help decompose the clippings and encourage microbial action. In one year, you will have a free, and environmentally friendly, source of compost.

If you do not have room for a compost bin or pit, check with your city to see if it has a composting site. Do not put clippings in with your household waste; it is illegal in some states and environmentally unfriendly everywhere. You can take your yard waste to the site and return later to pick up ready-made compost. The cost in my area is free — in fact, here they are begging people to come and pick it up. Elsewhere, it runs from $10 to $20 per truckload.

Composting

Compost is the most beneficial soil additive you can use, providing moisture in dry soils, drainage in heavy soils, and pH balance everywhere.

Compost bins are easy to construct, and you can find instructions online. If you do not have the time or patience, you can buy one readymade. They cost from as little as $30 for a small container to as much as $500 for a large, elaborate barrel mounted on a frame that you can spin by hand.

Composting bins use the heat of the sun, water, and sometimes a biological booster to rot vegetation into fine, dark, friable soil. Almost everything green goes in it — including things that were once green. You can add grass clippings, chopped tree branches, shrub prunings, wood ash, cardboard, paper, and vegetable waste, such as the tops of carrots. Leaves make fine compost, though the addition of oak leaves makes the compost somewhat acidic. Weeds can be composted, but the seeds might survive, even at temperatures in excess of 130 degrees, which is the desirable inside temperature of a compost bin. You can also add "green" food waste, such as leftover salad or vegetables, preferably without the salad dressing or butter. You can add tea bags, coffee grounds, and eggshells, and all will enrich the compost. As a rule, you want to keep the ratio of brown, like tree branches, to green clippings at 25:1 to keep the pile cooking. Smaller objects decompose faster than larger ones, so invest in a shredder.

Two of the most important elements in composting are air and water. If your compost bin is a roll of fencing, you are in good shape; wind and rain will do much of the work for you, though you may have to add a plastic tarp to get the sun to do its part. If you are thinking of buying a barrel composter, make sure it has side vents, and add water occasionally. Compost has precise moisture requirements. A handful should have about the same dampness as a wrung-out washcloth.

Plant pH is another vital consideration in compost. Pine needles and oak leaves are acidic, but hay and grass are alkaline. Try to keep pH materials in balance, or — if you cannot — test compost before using it as a soil amendment. Oak-leaf compost on a strawberry bed will produce small, bitter fruit. Compost made primarily of grass clippings spread around rhododendrons or azaleas will mean smaller blooms and unhealthy-looking plants.

Compost works best at about 130 degrees. If you do not know how hot your compost is, you can buy a thermometer. You can add a biological booster if your compost is less than 100 degrees, but read the instructions; like pH, the adjustments must be incremental. Compost bins rely on the sun's heat, so place yours accordingly. If your compost is getting too hot, you can cool it with water, or open the door to allow heat to escape. Keeping your compost at more than 150 degrees for any length of time will kill the bacteria, fungi, worms, and insects that make waste into compost.

As the population multiplies and waste management companies reach their landfill capacity, it becomes increasingly important to recycle yard and household waste. Composting fills this need. In return for a minor investment of money and time, nature will reward you with an unparalleled source of nutrition for your shrubs, flowers, fruits, and vegetables. Making compost is a win-win situation. For more information, go to **www.compostguide.com**.

Pesticides

There are no "pests" in a garden, merely inhabitants who do not look, think, or eat like you do. A garden, properly managed, will not have much burden of pests; these come from improper watering, lack of appropriate soil nutrients, and poor management.

Plant pests can be separated into three groups: sucking insects, leaf-chewing insects, and boring insects. In the first group are aphids, or plant lice, and these can be destructive. There are 4,000 species of aphids, of which 250 have earned most-wanted posters. Aphids can be white or greenish-white and less commonly pinkish or pale brown, depending on how old they are. Aphids often have lacy wings. They are found in colonies, often with ants, which farm them the way we raise cows, and, like us, for aphids' milk, which is honeydew. If you want to get rid of them, blend a habanero pepper in one cup of water, mix with two drops of Volck oil, available at any garden center, and spray the affected plants.

Spider mites are another sucking insect. They are so tiny they can look like reddish powder on the undersides of leaves, and a badly infested plant will have actual webs. They like tomatoes, peppers, and roses about equally, as well as 150 other cultivated species. When they wreak their havoc, the leaves of affected plants will look mottled, or burnt. Try mixing one tablespoon of buttermilk, one cup of unbleached flour, and four cups of water and spraying the mixture on affected plants. For mites and scale insects, you can soak a pack of cigarettes in a gallon of water, add a few drops of Ivory liquid soap, and use that. Be careful of children or pets; nicotine in concentration is poison. Fish-emulsion fertilizer also will discourage most nuisances.

Leaf-sucking insects such as caterpillars, beetles, leaf miners, and sawflies can turn a leaf into lacy shreds in a matter of hours. We all know what a caterpillar looks like. What not everyone knows is that some caterpillars turn into wasps. Wasps, though ugly and scary, benefit a garden by eating other pests. Once a caterpillar has rolled into its leaf, spraying does not do much good; these leaf-balls are solidly constructed and resist penetration. You can pick the damaged leaves off a plant and burn them. It is time consuming but the most effective way to get rid of caterpillars.

Leaf miners chew the edges of leaves. Sawflies masquerade as caterpillars, lay eggs, and ravage roses. The only way to tell a caterpillar from a sawfly is the legs. Sawfly larvae have — in addition to real legs — body bumps that look like undeveloped legs. Sawflies also have an effective natural predator in the paper wasp, but because of pesticide overuse, paper wasps have become increasingly rare. Try mixing a strong solution of one handful of garlic and a teaspoon of Ivory soap in a blender and diluting with water in a five-gallon sprayer.

Boring insects include the oak-leaf roller, birch borer, and elm borer. These insects can kill a tree. Their presence was less a menace in the early part of the last century, but increasing use of pesticides and urban sprawl have decimated woodpecker populations — the borer's only natural predator. The ivory-billed woodpecker went extinct about 1950, and only one has been sighted since. The red-cockaded woodpecker, also native to the Southeast, was listed as endangered in 1970.

Fungi — the plural of fungus — are distinguished from bacteria by their minute size and the fact that they have cells with discernible organs, such as nuclei and mitochondria. Fungi are divided into three categories: yeasts, molds, and mushrooms. All thrive in damp conditions. Drying out your garden will not kill them, because fungi have a remarkable ability to survive drought in spore form. They are a serious plant problem, but they also have a benevolent side. In the roots of plants, they form mycorrhizal, or symbiotic, relationships, helping plants take up nutrients from the soil and defending against other harmful fungi and insects. This makes them useful as biological weapons in large-scale gardening, such as greenhouses, forestry, and food production. Unfortunately, chemical fungicides wipe out both bad and good fungi.

Mildew is a form of mold. Most gardeners see it as either powdery mildew, on roses, or downy mildew, commonly found on grapes and vine-like vegetables. Powdery mildew is an obligate parasite, like a virus, and it cannot survive without a host. This is not useful knowledge to rose growers, who are not likely to uproot and burn a classic Therese Bugnet rose just to get rid of mildew.

There are some safe fungicides — neem oil, for instance — sold as organic fungicides, and organic gardeners swear by them. They are more labor intensive and not as immediately effective, but they also will not kill you, your family, your pets, or the honeybees that pollinate your flowers and food crops. You also can mix one part unpasteurized cow's milk, or one part strong coffee, in ten parts water. Alternatively, you can use baking soda, at one tablespoon per gallon, or vinegar, at one pint per gallon, depending on the pH of your soil. Add a drop of Ivory soap and a tablespoon of Volck oil to either mixture to bind the spray to the leaves. The latter two work because pH is a significant factor in plant disease.

If you must use commercial insecticides, use them seldom and wisely. Choose soap- or oil-based products, as they will be less harmful and remain on the leaves longer. Mix according to the manufacturer's recommendations, always wear protective gear, cover exposed skin, and dispose of containers properly. Insecticides are poisons.

Green Venture has kindly provided the following Case Study on how one neighborhood reduced the use of pesticides.

CASE STUDY: GLEN MARSHALL

Hamilton Coalition on Pesticide Issues (HCPI)

Glen Marshall

CLASSIFIED CASE STUDIES™ directly from the experts

Three years ago, we began a campaign in our neighborhood to reduce the use of pesticides. We live in the Southam neighborhood, west of Upper James St. and north of Fennell Avenue in Hamilton, Ontario, Canada.

We began this campaign for several reasons. My wife is a reporter who was assigned to do a series on cosmetic pesticides and became alarmed by the research she did. Shortly afterward we had a son, and soon after that, doctors released more information saying they believed these chemicals were harmful, particularly to children.

CASE STUDY: GLEN MARSHALL

The first year we simply sent out a letter stating our concerns and attached a report by the Ontario College of family physicians that outlined the health risks. In subsequent years, we have sent similar letters in early spring, just about the time the pesticide companies start making their calls. This last year, however, our communication expanded to include a newsletter. We discovered that it's not enough to tell people, "Do not do it." You have to tell them what else they can do and list some of the alternatives.

For the most part, our neighbors have been receptive to our efforts. We estimate that approximately 8-10 percent of homeowners in our immediate area still spray, down from approximately 28 percent in previous years. We do still have a few "vocal" holdouts, who have made it clear they have no intention of switching. We make it just as clear, however, that we have no intention of stopping our yearly campaign. We understand that people have a right to make their properties tidy, but we strongly believe this must not happen at the possible expense of another person's health.

Recently, we have expanded our efforts to include some nearby properties such as Mohawk College. With the help and guidance of Green Venture, a pilot project is now being planned to demonstrate the effectiveness of organic lawn care on institutional grounds. Along with Hillfield (which already uses organics) and St. Joe's Healthcare (which has stopped spraying as well), acres of nearby land will now be chemical free.

This year, we were happy to receive recognition for our work in the form of an award from the Halton Conservation Authority. We have also been asked to make a presentation of our efforts at Hamilton City Council. Please feel free to use the copies of our letters (which are included on this Web site) for circulation in your own neighbourhood. Good luck with your efforts!

Courtesy Green Ventures, HCPI, and Glenn Marshall

www.greenventure.ca

What is pH?

The term pH is a measure of the alkalinity or acidity of soil — as measured by the activity of negative hydrogen ions in solution. From zero to seven, soil is acid, or "sour." From seven to fourteen, soil is alkaline, or "sweet." A neutral pH is seven. A one-unit change in pH represents a tenfold change in hydrogen activity, so make changes slowly. Most soils, particularly heavy or uncultivated soils, or soils in damp climates, tend to be acidic.

There are three key elements in soil — nitrogen, phosphorus, and potassium. Nitrogen is an inert element. By a process known as microbial conversion, plants change nitrogen to nitrates, which they can assimilate. At a pH of seven, this conversion is rapid, and plants thrive. Below seven, the conversion slows. Phosphorus also is highly pH dependent. In alkaline soils, phosphorus is taken up in small amounts and bonds with calcium but is released relatively quickly. In acidic soils, phosphorus binds to iron and aluminum salts, and the bonding tends to be permanent. Potassium acts the same way, that is, available in alkaline soils but bound by aluminum in acid soils. Calcium and magnesium are readily available in neutral or alkaline soils. Sulfur, a micronutrient, is almost unaffected by pH, but manganese, iron, copper, zinc, and boron are higher in acidic soils and lower in alkaline soils. Soil pH does not just affect the availability of nutrients; an incorrect pH can encourage diseases by weakening a plant's immune system, just as stress weakens the human immune system.

Most vegetables prefer a neutral or slightly acid soil, with a pH from four to seven. Most trees, shrubs, and ornamentals do well in a pH range from six to eight, but there are exceptions. Hydrangea and spruce like acidic soil, at or below a pH of four. Azaleas, rhododendrons, and lily of the valley prefer a pH of five or six, as do dogwood and gentians. Daylilies, forget-me-nots, and wisteria will do well in alkaline soils, and most berries prefer sweet soil.

You can test your soil's pH with a kit available at any good garden store. Changing your soil's pH should be done slowly and in small increments; the microbial conversion of chemicals takes time. Lime makes soil sweeter. Four ounces per square yard in sandy soil will increase pH by one, and one is the increment you should always adjust by, repeating applications no more than twice per year. In clay soil, you will need 12 ounces per square yard. Loamy soil, which is made up equally of sand, silt, and clay, will need from six to nine ounces, and peaty soil, made up of millennia of rotted, compacted vegetation, might need as much as 24 ounces per square yard. Bone meal, ashes, and crushed oyster shells will also raise pH. To lower pH, you can use sulfur per the manufacturer's directions, or you can incorporate peat moss into the soil. You can also use sawdust, wood chips, or leaf compost — particularly oak leaves.

Lawn mowers

A new report from Sweden, published by Mindfully.org, shows that cutting grass for one hour creates as much air pollution as driving your car 100 miles. The report, which the authors say is the first to compare lawn-mower pollution with auto mileage, recommends using catalytic converters on mowers.

The EPA already has acted to cut emissions from lawn mowers, leaf blowers, and other small, gasoline-powered equipment by requiring manufacturers to install catalytic converters on all engines less than 25 horsepower. The rule will be implemented nationwide in 2011; California's emissions standards began in 2008. The EPA also will place new emissions standards on inboard and outboard engines for boats beginning in 2009. The EPA said these new emissions standards, which will raise small-engine prices from 5 to 11 percent, will reduce their emissions by 70 percent, or between 10 and 25 percent of all mobile-source hydrocarbon emissions.

You could one-up the EPA and put a converter on your lawn mower, if you could find one. I tried and came up with nothing; presumably, small-engine manufacturers are behind the curve, environmentally speaking. However, even if you did, the horsepower robbed from your lawn-mower engine by the converter would mean you would have to run it almost twice as long, defeating the purpose. This loss of horsepower when using a catalytic converter may be why auto emissions standards keep falling behind. Catalytic converters, while reducing primary emissions such as carbon dioxide, contribute nitrogen oxide to the atmosphere, and nitrogen oxide is a potent greenhouse gas, accounting for 7 percent of the greenhouse effect.

A better method for reducing air pollution would be an electric lawn mower. I have owned them and they work remarkably well. However, the best footprint-reduction strategy involves an old-fashioned, reel-type mower. You can still find them in hardware stores, second-hand stores, and occasionally, on CraigsList.

The best idea would be to reconsider our life styles and our need for an acre of grass. If we knew the real cost of that acre of grass, both financially and ecologically, we might grow to prefer wildflowers.

Cleaning Products: Clean but Deadly

Francis Bacon is supposedly the one who wrote, "Cleanliness is next to Godliness." As much as I admire Bacon, he did not live in the 20th century, with chemical pollution rising so fast scientists are no longer able to extrapolate its ultimate effects on the natural world.

According to Linda Maher in *Natural Health & Vegetarian Life*, winter 2007, there are about 100,000 chemicals on the market that simplify our lives. Unfortunately, many of these supposedly beneficial chemicals produce unnatural estrogens, called xenoestrogens, which disrupt the endocrine system. The endocrine system mediates all other bodily systems through the production of hormones.

Unlike natural estrogen, which is eliminated by metabolism fairly quickly, these foreign estrogens collect in fat tissue, such as breast tissue. What does this mean to our bodies? Breast cancer rates have doubled since 1970, young girls begin their menstrual cycles earlier, effectively raising their likelihood of breast cancer by 25 percent, and women chemists and hairdressers have a 30 percent higher risk of developing breast cancer. According to the United States Centers for Disease Control, prostate cancer rose 300 percent from 1973 to 1992, and the death rate from prostate cancer rose 25 percent. Birth rates are declining, hitting an all-time low in 2002 according to another Centers for Disease Control report, and fertility clinics report business is skyrocketing. Polycystic ovarian syndrome is on the rise, as are uterine fibroid tumors; almost 50 percent of women have them, and 500,000 hysterectomies are performed every year to remove them, according to Dr. Elizabeth Smith, who maintains a Web site called Fibroid 101(**www.fibroid101.com**). *The New England Journal of Medicine* reported in February 1995 that sperm count has declined 33 percent during the past 20 years among a study population of 1,351 healthy, fertile men in Paris, France. People who work — or worked — in the plastics industry have a greater risk of developing brain cancer, according to a December 2006 article in *U.S. News and World Report.*

Many of us have learned, when buying food, to reject anything whose ingredients cannot be pronounced or identified. This option does not exist for household cleaning supplies or cosmetics, some of whose ingredients are proprietary. The FDA requires cleaning agents to contain warnings if one of the ingredients is hazardous, and the danger, whether toxic, flammable, or carcinogenic, has to be identified. Otherwise, the burden is on the manufacturer to do a toxicological analysis. However, no law requires a full list of ingredients.

A Primer on Persistent Organic Pollutants

Though the term persistent organic pollutants (POPs) is a relatively new one, the chemicals that it encompasses are not. Instead, what is notable about this new category of pollutant is the fact that for the first time the many existing toxic compounds it includes have been grouped together under a single label. This shift marks an important change in the way we think about chemicals, the pollution they create, and the ways in which we might regulate them.

Since synthetic chemicals first began to be mass produced in the early 20th century, they have been largely regulated on a case-by-case basis. When there were only a handful of compounds being manufactured in small amounts, this piecemeal approach worked, but the swift growth of the chemical industry and the avalanche of new materials it unleashed soon rendered this strategy ineffective. According to a paper published by the American Chemical Society, in 1997, there were more than 70,000 different chemical compounds in production and some 6 trillion total pounds of chemicals being manufactured in the United States each year. Yet in spite of this variety and volume, an obsolete, one-chemical-at-a-time approach to regulation continues.

Creating a new category of chemicals based on these persistent organic pollutants represents the critical first step toward developing a better, more-effective regulatory system, one based not on controlling chemicals individually but on regulating entire classes of compounds with a single set of rules. The persistent organic pollutants category takes this approach one step further by using a new and different set of parameters to determine whether to assign this label to a particular chemical. Previously, science and government tended to group compounds together according to their chemical similarities. Membership in the persistent organic pollutants family of chemicals, on the other hand, is determined by how a specific chemical behaves in the environment and in the human body.

Persistent organic pollutants include many pesticides, industrial chemicals such as polychlorinated biphenyls (PCBs), organochlorines, and by-products of a variety of manufacturing and waste-incineration processes, such as dioxins. Because it is a new kind of chemical category based on health and environmental effects, not chemistry, any compound can be labeled a POP, as long as it has these characteristics:

- It resists biodegradation and therefore persists in the environment.
- It builds up in body fat and accumulates in ever higher levels as it migrates up the food chain.

- It travels efficiently throughout the atmosphere and global waters.
- Many POPs are linked to serious hormonal, reproductive, neurological, and immune disorders.

(Courtesy Seventh Generation: **www.seventhgeneration.com/about**)

Makeup, Cosmetics, and other Personal-Care Products

Canada is one step ahead of the United States. In November 2006, Canada required all cosmetics manufacturers to post a full list of ingredients on the packaging. The United States cosmetics industry is scrambling to comply, because 70 percent of all cosmetics sold in Canada are manufactured in the United States or by American firms outsourced overseas.

A study done in Great Britain a few years ago shows that aerosol air fresheners caused a significant increase in postnatal depression among women and a 30 percent increase in earaches among infants six months or younger. Aerosols also increased infant diarrhea by 22 percent. The biological mechanism involved in these illnesses is not fully understood, but it is suspected that the chemicals in air fresheners weaken the body's defenses by making the skin more permeable. Most air fresheners tested in the United States contained varying degrees of phthalates, according to a 2007 report by the Natural Resources Defense Council, and these chemicals are a known or suspected cause of everything from uterine fibroid tumors to sterility and cancer.

A More Natural Approach

Not all "natural" products are natural. Some manufacturers, who describe their products as "natural" or "environmentally friendly" use anionic and non-ionic surfactants in their laundry soaps, and their manufacture creates toxins. Anionic/non-ionic surfactants are alkyl benzene sulfonates (ABSs). Non-ionic surfactants, or linear alkyl sodium sulfonates, are called LASs. ABS compounds are seldom used anymore, but LAS compounds are still fairly common, in spite of oleo-based substitutes. During manufacturing, linear alkyl sodium sulfonates release benzene, a reproductive toxin, into the air. They biodegrade slowly and are irritants when diluted but toxins in greater doses. The problem with all environmental toxins is cumulative: one milligram per month will not kill you, but one milligram of each of 1,000 toxins over a lifetime will.

There are hundreds of companies in dozens of cities creating the same kinds of threats to our health and our environment. Their footprint, though not always carbonized, is huge. Some sell "green" soap, others "gold;" the net effect is the same — poison that accumulates in our bodies slowly and manifests as reproductive anomalies or cancer.

Seventh Generation, which manufactures a line of all-natural cleaning and household products sold in retail stores, is an excellent resource for identifying some common toxic chemicals in our homes. Geoff Davis, Seventh Generation's in-house writer and media representative, provided the next Case Study.

CASE STUDY: SEVENTH GENERATION

Seventh Generation

Geoff Davis

As promised, here are some thoughts about conventional cleaning products (i.e., those that are made from synthetic chemical compounds and surfactants) as they relate to both a healthy home and indoor air quality. It's quite likely more than you need, but I thought it better to provide too much rather than not enough given how important we feel these issues are.

CASE STUDY: SEVENTH GENERATION

There are a wide variety of health issues at play where conventional cleaners are concerned. They begin with the ingredients typically found in these products. Conventional cleaners are largely made from synthetic chemicals, which are themselves largely derived from petroleum, a non-renewable, unsustainable resource whose extraction, transportation, and refining harm the environment in many ways. While this aspect of the situation does not directly impact the state of one's home, we feel that it is nonetheless a critical part of the overall picture. We think that the term "healthy home" is about more than simply what's going between the four walls of one's house. We think it should extend out into the larger world as well to encompass the general environment. In this view, establishing a healthy home is not just about protecting the health of its occupants and their immediate living spaces. It is also very much about adopting practices and strategies that protect the local, regional, and global biosphere as well.

The decision to steer clear of conventional cleaning products contributes to this larger picture in other ways, too. When we use these products in our homes, their remains are often rinsed down our drains. If we have a septic system, this means the chemicals conventional cleaners contain will eventually filter out into our local soils. If we rely on a centralized municipal treatment plant, it is very likely that the chemicals will end up in those nearby waterways that receive the plant's treated water. This is because our country's sewage treatment infrastructure is badly aged and uses outdated technologies that are unable to process modern water contaminants, including cleaning product chemicals. Whether we use a septic tank or are connected to a larger wastewater treatment system, the chemicals we put there eventually make their way into the food chain, where they often persist rather than biodegrade harmlessly.

Inside the home there are a host of issues that synthetic chemical cleaning products raise. These start with two important points that need to be made about such products:

1. There are no laws that require cleaning product companies to submit their products to third parties for unbiased safety testing; neither is federal approval of a product formula required before it can be sold. (Disinfectants and sanitizers are a notable exception; they are regulated by the EPA as pesticides and subject to prior approval.) Regulators at agencies such as the Consumer Product Safety Commission play little or no role in reviewing test results and approving a cleaning product for sale in the U.S. In place of strict product testing and government oversight is a system of largely voluntary standards in which manufacturers themselves are responsible for all initial assurances of their products' safety. Cleaning product companies are legally free to sell almost any product with almost any ingredient and need little more than their own okay to bring it to the market. Instead of a precautionary look-before-you-leap

CASE STUDY: SEVENTH GENERATION

approach, it is "shoot first and ask questions later" where cleaning products are concerned. In most cases, any hazards or health effects that are caused by exposure to a particular cleaner or ingredient become known only after people have used the product in question, gotten sick, and reported this fact to federal authorities who only then are permitted by law to take action. Unfortunately, by the time such complaints usually surface, the product or ingredient causing all the trouble has been in widespread use for months or even years.

2. The government does not require that manufacturers disclose all the ingredients a given product contains on its label. The regulatory situation here is a bit complicated, but here's the rundown:

The Federal Hazardous Substances Labeling Act of 1960, the set of laws that govern today's cleaners, requires only that manufacturers list "the common or usual or chemical name of the hazardous substance, or of each component, which contributes substantially to its hazard" (see below). In other words, if an ingredient is not connected to a product's known acute or chronic hazards, it does not have to be listed on the label. Manufacturers are legally permitted to withhold any and all information about the bulk of the different chemical compounds they use in their cleaners. These secret ingredients can include all kinds of substances from solvents, dispersal agents, and carriers to preservatives, buffering agents, and "inert" ingredients.

When it comes to these hazards, cleaning product labels must warn consumers about any "immediate" dangers (i.e., acute hazards) that may occur in the event a product is used incorrectly. This required warning takes the form of "signal" words such as DANGER or POISON and WARNING or CAUTION.

Cleaning products are also regulated under Title 16, Chapter II of the Code of Federal Regulations. Title 16 expands the definition of the word "toxic" as it applies to cleaning products. This revised definition addresses chronic health effects and declares that any ingredient that is a confirmed or probable carcinogen, neurotoxin, or developmental or reproductive toxin must be considered "toxic" as well and subject to the same labeling regulations as ingredients that pose immediate acute hazards. A cleaning product that contains any ingredients that meet this expanded definition would also have to be labeled with a signal word. Unfortunately, a loophole allows manufacturers to combine their warnings about different hazards into a single statement if that statement gives consumers all the information they need to deal safely with each individual danger. An even bigger loophole in labeling laws means that in many cases cleaning products do not even have to do this much. According to current regulations, a manufacturer can use a substance known to be chronically hazardous but omit all warnings or references to it by claiming that any exposure to that substance resulting

CASE STUDY: SEVENTH GENERATION

from the product's use would not be not large enough to trigger the toxic effect. Manufacturers also are allowed to figure out for themselves how to take the animal studies that typically demonstrate chronic risks and extrapolate them to human beings for such risk assessment purposes. These loopholes essentially let manufacturers decide for themselves what's hazardous (and therefore what requires disclosure to consumers) and what is not, and this, in turn, allows them wide latitude in communicating information about potential chemical hazards. The end result is misleading labeling at best and deceptive labeling at worst.

In all cases, consumers lose. They have no real way to tell what hidden hazards they're bringing into their homes when they buy a given conventional cleaning product, and lack of required safety testing means they could be bringing home almost anything.

Indeed, what we do know from various independent testing and analysis is that conventional cleaning products contain all kinds of unhealthy and outright dangerous substances. These include carcinogens, neurotoxins, reproductive and developmental poisons, hormone-disrupting compounds (substances that mimic hormones in the bloodstream to deleterious effect), organ toxicants, and immune system suppressors.

Independent research by concerned organizations is uncovering new hazards all the time. Recently, for example, the Montana-based nonprofit Women's Voices for the Earth examined Material Safety Data Sheets for dozens of top-selling conventional cleaning products and found that most of formulas they studied contain one or more chemicals linked to asthma and reproductive disorders, including monoethanolamine (MEA), a solvent that appeared in laundry detergents and all-purpose and floor cleaners; ammonium quaternary compounds, disinfectants used in disinfectant sprays and toilet cleaners; glycol ethers, solvents added to products such as glass cleaners and all-purpose spray cleaners; alkyl phenol ethoxylates (APEs), surfactants used in laundry detergents, stain removers, and all-purpose cleaners; and phthalates, fragrance carriers that were discovered in glass cleaners, deodorizers, laundry detergents, and fabric softeners.

These hazards can come from any number of different ingredient types, including surfactants (the "active" ingredients that actually do the cleaning) and the so-called "inert" ingredients that typically make up the bulk of a cleaner's formula. The "inert" classification, however, is highly misleading because so-called inerts are often anything but. They're labeled "inert" not because they're harmless but simply because they are not technically considered an active ingredient involved in the marketed function of the product. (Household chemical products generally consist of active ingredients, such as the chemicals that actually do the work, and inert ingredients, a catch-all term for all the remaining "non-active" ingredients.) Inert ingredients can include

CASE STUDY: SEVENTH GENERATION

buffering agents, solvents, preservatives, dispersal agents and carriers, wetting agents, fillers, and other ingredients that help stabilize, dispense, and increase the potency, effectiveness, and ease of use of the product. All of these can be and often are made from toxic chemicals.

When we use these products in our homes, our families are exposed to these toxins in a variety of ways both during and after the actual use of the product.

It is important to keep in mind that we are talking about exposures that consumers cannot generally see or feel and so usually have no idea is occurring. These contacts are generally happening on a molecular level, which is where all the biological "action" that we are concerned about takes place. Except during the relatively few and far between moments when a given product is actually being used or applied, rarely does a person encounter a cloud of cleaner vapor or a puddle of cleaner on a counter. Instead, we are usually (and perhaps more dangerously) exposed to small amounts of cleaning product chemicals after the fact. These contacts may have little individual impact (though not always — it depends on the chemical in question) but can add up over time to create a chronic exposure that can easily cause a disease or other negative health effect.

Though our families' exposure to the hazardous substances contained in conventional cleaners is obviously greatest when the product is being used, it does not end there. There are two ways that cleaners affect indoor air quality. Both can spread the contamination that cleaning products create well beyond the immediate area actually being cleaned and well past the time this cleaning occurred. Here are the primary routes by which we are exposed to the hazardous chemicals conventional cleaners contain:

1. Many products leave residues behind on the surfaces to which they've been applied. This is a fairly straightforward transmission route. We apply a product and wipe it off to remove dirt and soils, but significant traces invariably remain. These traces can be absorbed into our skin when we touch the surface in question or ingested when we transfer them from our hands to food or drink. Babies and toddlers are especially vulnerable to this type of transmission since they are always putting things, including their fingers, in their mouths.

2. Many products release vapors when they are used. Cleaners frequently contain volatile organic compounds (VOCs) in the form of solvents to help remove grease and oils and drying agents to speed evaporation. (Though these are by no means the only purposes VOCs can serve in a cleaner.) These ingredients readily turn to vapors at room temperature, which are then inhaled by anyone in the home for hours and even days after the product has been used. Vapors also drift on air currents and can travel around the home to distant points.

CASE STUDY: SEVENTH GENERATION

Vapors can also be released by using cleaners in or with hot water, which accelerates any vaporization. Much of the chlorine in conventional dishwashing detergents, for example, is vaporized by hot dishwasher water and released into a home's air during the wash cycle or when the dishwasher is opened too soon after the cycle completes.

3. Spray cleaners form aerosols when they are used. These are microscopic droplets of product that are so tiny and lightweight they can remain suspended in the air for hours after the product has been used. Aerosols are easily inhaled, and they can drift around the home and come to rest on food, clothing, and other surfaces far from the place the product was applied.

These are the basic concerns surrounding conventional cleaning products. There are lots of different tangents involving the specific behaviors of specific ingredients in the home, the human body (especially children's bodies), and the environment, but that's the basic gist of the situation. Manufacturers can and do commonly use very hazardous chemicals in their conventional cleaning product formulas. They do not test or even tell consumers that these ingredients are present. And when we use these products in our home, we contaminate indoor air and surfaces with toxins that we then ingest, inhale, or absorb and that can cause negative and even deadly health effects when we do.

There are some other concerns I should mention in passing:

1. Many of the chemicals we're talking about are persistent. They do not readily break down into less harmful component parts but instead tend to remain intact and therefore biologically active for long periods of time. Some compounds in cleaners can literally last for years. This is especially true of chlorinated hydrocarbons, substances made by combining chlorine with hydrocarbons from petroleum. Once they're set loose, persistent chemicals tend to remain for a while.

2. When we ingest persistent chemicals, they tend to accumulate over time in our body tissues. Many persistent chemicals are fat-soluble, which means they are stored in fatty tissues rather than excreted by the body. This results in something called our body burden, the total amount of persistent chemical pollutants absorbed and/or ingested by our body over time and semipermanently stored in its tissues.

3. There is also the issue of synthesis, which occurs when the chemicals in the different products we use around the house come into inadvertent contact with each other and accidentally combine to create brand-new compounds. These new substances can sometimes be more toxic than any of the original chemicals. For example, many people are surprised to learn that combining

CASE STUDY: SEVENTH GENERATION

ordinary household chlorine bleach with ammonia will cause these two chemicals to immediately react with each other and produce a deadly gas. Reactions like these between any number of the compounds found in the many products we use have great potential to create new hazards. In most cases we cannot see the uncontrolled chemical reactions that take place during accidental synthesis, nor are there any clues that something new and potentially toxic has been created.

4. Interactions between the chemicals found in different products can sometimes enhance the potential for harm of any or all of the individual chemicals involved. This process is known as potentiation. For example, the solvent acetone, which is found in products such as nail polish removers and waxes, has been shown to increase the liver damage caused by carbon tetrachloride (a household chemical now thankfully banned), even though acetone itself does not harm the liver.

For all these reasons, we do not think conventional cleaning products have any place in a healthy home. A precautionary approach, which is one we always recommend, dictates that all conventional cleaners be avoided until we know exactly what is in each individual formula and can ascertain with certainty the level of danger those specific ingredients represent. It is our fundamental belief that we just do not know enough about the chemicals and cleaning products that surround us to continue using them. In fact, the available evidence strongly suggests that these materials are causing us harm. Until we know what harm is being caused by which chemicals and which chemicals are, by extension, safe, we believe consumers should stop using all of them. We think that is just simple common sense, especially when you consider that the use of conventional cleaning products is wholly unnecessary. Consumers can easily make their own homemade cleaners using common, inexpensive, nontoxic ingredients such as vinegar, essential oils, baking soda, and washing soda. Or they can purchase more convenient pre-made nontoxic cleaners like our own Seventh Generation products, which are made from safe and natural biodegradable ingredients like vegetable oils and harmless hydrogen peroxide.

Last, I would like to recommend two additional resources to you. The first is our book, *Naturally Clean*, which I co-wrote with Seventh Generation president Jeffrey Hollender. It represents the basic sum of our knowledge on the subject and is the culmination of years of educating ourselves and our customers about these issues.

Much of what I have said here is drawn from this work, and you'll find it expands on the subjects I have discussed to provide additional useful information, evidence, scientific studies, references, etc. This book was written in the second half of 2005. To catch up on all the research and related developments that have occurred since that time, I

CASE STUDY: SEVENTH GENERATION

would like to refer you to the online archive of the *Nontoxic Times*, our free consumer newsletter. You can find the archive at: **www.seventhgeneration.com/learn**.

Perusing the issues of our newsletter that we've published since completing the book (roughly the last two years' worth) will bring you up to date on the state of the things and provide additional emerging evidence that we should be very, very cautious about using conventional cleaning products. Public education is a key part of our mission, and we are always looking for and appreciative of any ways we can alert consumers to our publications and the important messages they deliver.

A third, and perhaps the "grandfather" of the green cleaning movement, is Simple Green. In business for more than 30 years, Simple Green offers more than 20 different household and industrial-strength cleaning products that are nontoxic, biodegradable, and formulated to tackle even the messiest spills and stains. Simple Green even provides Material Safety Data Sheets on its Web site, for customers who want to compare before they buy. You can buy Simple Green products at hundreds of locations, including Ace Hardware, Lowe's, and Menards. The following Case Study, given to me by Carol Chapin, Environmental and Regulatory Director, and written by Dr. John Todhunter, former head of the EPA's Office of Prevention, Pesticide, and Toxic Substances, represents Simple Green's position on green cleaning.

CASE STUDY: SIMPLE GREEN

Simple Green

Why Simple Green in 1975, and Why, Still, Simple Green Today?

These days, so many products are screaming out, "I'm green!" They wrap themselves in the flag of this, that, or the other "natural," "organic," or "green" certification. What they do not tell you is that getting these "seals of approval" is as simple as using only the ingredients that are on a certifying group's approved list of ingredients. That is, most of these products are not actually tested to see if they really are safe for health or for the environment!

CASE STUDY: SIMPLE GREEN

Of course, you would think that if only "approved" ingredients are used, a product should be "all right." This is not always true. The safety of multicomponent products, such as cleaning products, depends to some degree on what is in them but even more on how much of each ingredient and the specific combination of ingredients in the product. Too much of a good thing, or in the wrong combination with some other ingredient, can lead to unexpected consequences. Only careful evaluation of a product's formula and a solid program of product testing can assure that a product really is as good as it claims.

Simple Green is one of the most, if not the most, tested cleaning products on earth. It has been extensively tested for safety to human and animal health and for effects on sea life and other aquatic life. As a result, Simple Green has been independently evaluated by scientific experts who have concluded that Simple Green can be labeled "nontoxic," "biodegradable," and "environmentally safe." I am one of the experts involved for years with testing and evaluating Simple Green's health and environmental safety. One typically does not find such claims made for other cleaning products because they cannot back them up. Simple Green's claims have held up before the Federal Trade Commission, which regulates advertising claims, and is in accordance with Consumer Product Safety Commission and EPA regulations and guidelines for defining these terms. That is a much higher standard than the simple, ingredient-list-only programs some other "green" products tout. Copies of the full scientific reports that summarize the hard evidence supporting Simple Green's claims can be obtained by contacting Sunshine Makers, Inc.

The next time someone touts a product with "green" certification, ask the right question: "What is it based on?" If it is not based on extensive product safety testing such as Simple Green's, then that claim is not worth much. You might ask, also, "Why does not Simple Green get one of these 'green' seals of approval?" The answer is simple: Simple Green does not want to be lumped in with other products under these simplistic systems. Simple Green wants the data to speak for itself and allow customers to choose between real testing/evaluation or a glossy seal. I think customers will want an assurance of safety backed up by real data. Simple Green has been tested or evaluated for the following types of health, environmental, and efficacy issues (This is only a partial list of the total testing that has occurred over 35-plus years):

- Oral toxicity (essentially nontoxic)
- Dermal toxicity (essentially nontoxic)
- Ocular toxicity (mildly irritating)
- Potential to cause cancer (essentially no risk)
- Potential to damage nervous system (essentially no risk)

CASE STUDY: SIMPLE GREEN

- Reproductivity and fertility (not a toxicant or inhibitor)
- Ames mutagenicity test (non-mutagenic)
- LD50 ranking (between ethanol and glycerin)

And more:

- Biodegradability (OECD 302B, ASTM D 2667-70, OECD 302A) (highly and completely biodegradable)
- Toxicity to freshwater fish (practically nontoxic per EPA rating)
- Toxicity to saltwater organisms (practically nontoxic per EPA rating)
- Viable bacteria testing (Simple Green does not contain viable bacteria)
- Volatile organic compounds (EPA Method 624)
- Volatile organic compounds (EPA Method 24)
- Semi-volatile organics (EPA Method 625)
- TCLP (EPA Method 1311)
- Volatile organics (EPA 8260, pass)
- Semi-volatile organics (EPA 8270, pass)
- Organochloride pesticides (EPA 8081, pass)
- Chlorinated herbicides (EPA 8151, pass)
- RCRA metals, pass

And more:

Aluminum cleaning, Aluminum cleaning before brazing, Anions by EPA 300.0 (chloride and fluoride), Bacteria: heterotropic plate count (APHA Standard), Bioremediation: effects on bacteria, Cleaning and wax removal (CID-A-A-39A), Cleaning ability (ASTM G121 & G122), Compressed as association cleaning efficacy factor, Cleaning in place for pharmaceutical use: total organic carbon/UV and Conductivity, Cleaning Efficacy (FSTM 536-6701), Cleaning Painted Surfaces (CSMA CDD-04), Conductivity by EPA Method 120.1, Deleterious effect on surfaces (FS PD 233-C), Firing Residue Removal (MIL-L-G3460D), Hyperbaric Chamber Cleaning — CHHI Cleaning Procedures H (Naval Sea Command), Materials Compatibility (GPU Nuclear & Naval Sea Command), Metals Analysis — ICAP Emission Spectrometry, Navsea 02 Cleaning, Optimum Cleaning

CASE STUDY: SIMPLE GREEN

Temperature Analysis, Oil Dispersant Testing per EPA National Contingency Plan, Phenols, PCBs & Chromium Testing, Precision Cleaning Testing (Research Triangle Institute), Soil Removal (FSTM 1747C), Specification Analysis (MIL-D-1679E), Stress Crazing of Acrylic Plastics (ASTM 484), Surface Tension (ASTM D1331-89), USDA Certification Testing (A4, A8, B2), Vapor Suppressant per AQMD Rule 1166, Window Cleaning (CSMA DCC-09, and more.

Of course, I believe the performance data noted above show that Simple Green is the best but gentlest all-purpose cleaner/degreaser on the market. Not only is Simple Green safety proven, but also it works.

With so many excellent choices in household cleaning products, I hope you will never again pull a can of air freshener off the shelf in your local grocery store, take it home, use it, and think you are doing your family a service by deodorizing the air. You are making them sick, even though the illnesses might take years, and perhaps generations, to manifest themselves.

Natural Foods

This book would not be complete without covering food. Food, and food growing, affects the environment. The food you consume affects you and your family. Food is as much a part of ecological survival, or impairment, as the products manufactured to build your house. In fact, without food, you would not need a house.

Genetically modified organisms (GMOs) are sweeping agriculture. The technology has so far been applied to corn, potatoes, and cotton. In Mexico, which bans the planting of genetically modified corn, native strains of true corn, called maize, already have begun to take on the genes and traits of genetically modified corn.

Each of these native strains has different defenses against bacteria and viruses, designed to prevent the complete eradication of the species. If all strains of maize become genetically modified, as some scientists predict, a single super-virus might adapt sufficiently to our arsenal of pesticides and the like and wipe out the species entirely. This will leave us without corn and without adequate genetic material to create more. Because corn is one of the main staples of agriculture, millions will starve. At this time, there appears to be no way to reverse these transgenic mutations. Genetically modified corn now accounts for 30 percent of all corn.

Genetically modified corn, such as Starlink, which is adapted to fight certain pests by using a Cry9c protein, was shown to produce allergic reactions in humans. Aventis voluntarily withdrew its license for Starlink corn in 2000, three years after the product had been introduced. Even today, there is no knowing what kind of genetic inroads Starlink might have made during its three-year debut in agriculture. Another genetically modified corn still on the market contains Bt10, a gene that helps the corn plant repel predators but is linked to a gene that, in humans, resists the effectiveness of a commonly used antibiotic, ampicillin.

Doctors first began to observe antibiotic-resistant strains of bacteria in the 1950s. Since then, the incidence of methicillin-resistant staphylococcus aureus (MRSA), campylobacter, streptococcus, and pneumococcus bacterias have risen exponentially. In the period from 1974 to 1997, the increase was a stunning 38 percent, according to the United States Centers for Disease Control. New antibiotics cannot be developed fast enough to conquer outbreaks. Vancomycin, formerly the drug of last resort, has been replaced by linezolid and the class of carbapenems.

According to a report sponsored by the United States Department of Health and Human Services and published by the National Institute of Allergy and Infectious Diseases, at least 70 percent of nosocomial infections, or infections acquired in a hospital, are resistant to the antibiotics normally used to treat them, requiring longer hospital stays and the use of more expensive drugs that might be more toxic. In a March 2004 *New York Times* article by Abigail Zuger, 20 to 30 percent of staph infections occurring among children in Miami and Los Angeles are antibiotic resistant. In Houston, these rates approach 50 percent. Some of the problem may be attributed to overuse of antibiotics among the population. Another factor might be an increase in ultraviolet radiation as a result of ozone depletion, which seriously affects our immune systems. However, genetically modified foods, particularly corn, also are culprits.

Unfortunately, you will not know whether you are eating genetically modified corn in a taco unless there is a product recall. Genetically modified corn is not approved for human consumption; nonetheless, a certain percentage of it escapes into the food supply every year, either by accident or by undetected transgenic modification. For more information on the effects of genetic modification of food, please read the Case Study supplied by Avant Gardening at the end of this section.

Pasteurized milk is another health concern. We have been pasteurizing milk in this country for almost 100 years, unwittingly depriving ourselves and our children of some of the most potent antibacterial and anti-inflammatory properties available from a whole food. Raw milk does not support the growth of diphtheria, staph, anthrax, or strep bacteria. These bacterial-inhibiting factors are lost when milk is heated to 180 degrees, which is what pasteurization does. Raw milk contains folic acid and vitamins B6 and C, which are destroyed by pasteurization. Raw milk also contains enzymes such as lactase, lipase, and phosphatase. Lactase helps digest the sugar lactose, and its absence in the diet is causally linked with rising rates of diabetes. Lipase digests fats, and its absence might be the culprit in high cholesterol. The lack of phosphatase, which helps absorb calcium, might link to osteoporosis. Raw milk also contains beneficial bacteria that aid in digestion and support the immune system, and cream contains a cortisone-like agent called myristoleic acid, which helps combat arthritis.

Pasteurized milk in this country is not only boiled beyond recognition, but also chemically enhanced with colors, flavors, and proprietary ingredients to such an extent that, by the time it gets to your table, it is little more than a white toxin. If we are worried about harmful bacteria entering the food chain through milk, we would do better to control the feed milk cows receive, which can include meat from other animals, even diseased animals, as well as feathers, hair, skin, blood, manure, antibiotics, and plastics. We would also improve the conditions under which milk is "harvested." Factory farms are the main culprits, and factory farming of sentient beings is reprehensible.

Another threat to our health and our environment is processed foods, or packaged meals. These foods, in boxes and/or bags, are a timesaver, but what is the good of saving 15 minutes on dinner if you are shortening your life by ten years? A good rule of thumb when buying packaged meals is that if there are more than ten ingredients on the box, and four of them are unpronounceable, you should not be eating them. Processed foods, such as ready-made frozen dinners, snack crackers, many cereals, packaged pasta concoctions, and the like contain two or three times the recommended daily allowance of sodium, set at about 2,000 milligrams, depending on a person's age and health.

These processed foods also contain additives and preservatives, which can cause everything from hives to asthma or even neurological complications. Sulfites are the primary culprits, but food colorings, gums, and antioxidants, such as BHA and BHT, also can trigger reactions, even in people who have ingested such substances many times previously. As any immunologist knows, allergies are both cumulative and incidental. In addition, the amount of energy required to produce and package these foods, the resulting packaging waste, and the dumping of excess ingredients or food wastes into landfills or water supplies is a further threat to the ecology. If you doubt this, visit a manufacturing plant and see the salt, grease, food additives, and food waste that are dumped into these quick but essentially unhealthy, foods.

Another health and environmental hazard is "fast" food. Fast-food chains buy factory-farmed meat, which furthers one of the most reprehensible industries in our modern world. They also buy the cheapest ingredients they can find so that profit margins remain high to benefit stockholders. There is no benefit to the consumer except instant gratification. These fast-food chains add fat and salt to their concoctions for flavor. Any American whose diet is one-third fast food is guaranteed to have all the precursors for diabetes and heart disease by the age of 30 and full-blown disease symptoms by the age of 50. Unfortunately, fast food is also affordable food if one does not have the money to buy food in quantity, the space to store it, or the facilities to cook it.

Remember when the industry told us how bad raw sugar was for us and introduced saccharin? The FDA has established that saccharin does not cause cancer, but that does not change the fact that it is manufactured from an O-toluenesulfonamide derivative and causes skin reactions and sensitivity to sulfonamides, a group of common antibiotics. Toluene ranks close to the top of the list of hazardous chemicals. Then came sucrose. Manufacturers claim that the chlorine in sucrose, or sucralose, is identical to the chlorine atom in salt, but it is more like a chlorinated pesticide. The ultimate effect on human tissue will not be known without long-term, independent research, but the FDA approves. Tagatose, another sugar substitute often used in diabetic candies, has been shown to cause gastrointestinal distress, including diarrhea, nausea, and flatulence. Apparently, it passes through the digestive tract without being absorbed. This is a novel and disturbing experience for the human intestinal tract, which is designed to extract a caloric value from everything it receives, and the symptoms may be a good indication that Tagatose is not a healthful food.

By now, everyone has probably read at least one report on the dangers of Aspartame. The newest kid on the block among nonnutritive, or sugar-free, sweeteners, Aspartame has been tested primarily on animals. It is a common ingredient in diet soda and may cause lung tumors, breast tumors, tumors of the endocrine system, several forms of leukemia, and even chronic respiratory disease. Reports are mixed, and highly conflicting, but for the moment Aspartame remains on the market, under FDA approval.

If the FDA is no longer protecting our food supply, we can protect ourselves by buying whole foods. Buy unpasteurized milk from a local dairy or from a whole-foods store. Buy honey to replace sugar. Honey, historically the first harvested sweetener, has been recognized for decades to inhibit a broad range of bacteria and fungus. It is naturally acidic, retarding spoilage, and contains an enzymatic form of hydrogen peroxide, which is widely used to curb infections. Honey also contains phytochemicals known as flavonoids, which have natural antioxidant and disease-fighting properties. Buy free-range meat, or substitute beans, seeds, or raw, unshelled nuts. Seeds, particularly sunflower seeds, contain pacifarins, which increase resistance to disease, and auxones, which help the body produce vitamins and rejuvenate cells, delaying aging. Unshelled nuts provide large amounts of protein, "good" fat in the form of high-density lipoprotein (HDL); as well as potassium; calcium; phosphorus; Vitamins A, B, C, and F; and other nutrients needed to support healthy brain function. Peanuts, which are a legume rather than a nut, also are high in vitamins and minerals, and the increasing incidence of allergies might be due more to immune system dysfunction than the legume itself, according to the Mayo Clinic's Web site. Peanut allergies are nonetheless dangerous, so eat them sparingly until you know how your body will react, and always be prepared for an allergic reaction, keeping in mind that allergic responses are cumulative.

When it comes to food, you should buy it fresh, preferably organic, and cook it yourself. A good diet would include 20 percent fruits and vegetables, 50 percent complex carbohydrates — as found in fruits, vegetables, whole grains, seeds, and nuts — 10 percent fat, 10 percent protein from meat or eggs, and 10 percent from dairy products such as unpasteurized milk, cheese, butter, cottage cheese, or yogurt. This is what our grandparents ate, and the incidence of cancer, diabetes, and heart disease in recent years indicates their diets were far more conducive to health and longevity than ours.

Avant Gardening supplied me with its viewpoint on genetically modified organisms, which is provided below:

CASE STUDY: AVANT GARDENING

Biodiversity and Genetic Engineering

Courtesy Avant Gardening: Creative Organic Gardening

www.avant-gardening.com

Genetic engineering is a process of artificially modifying plant or animal cells by cutting and splicing DNA from one cell into another for the purpose of transferring desirable qualities that will make a crop resistant to herbicides or insects or to enhance food value.

When genetic engineers insert a new gene into any organism, there are "position effects." These effects can lead to unpredictable changes in patterns of gene expressions and genetic functions. The protein product of the inserted gene might carry out unexpected reactions, producing potentially toxic products.

Living organisms are highly complex, and genetic engineers cannot predict all the effects of introducing these new genes. Problems that might develop from this process include new toxins and allergens, loss of bio-diversity in seed and crops, or damaging health effects from manipulated food crops. When new genetic information is introduced into plants, bacteria, insects, or animals, it can then be passed into related organisms through naturally occurring processes, such as cross-pollination.

It is estimated that 70 percent of the current genetically modified (GM) harvest is made up of herbicide-resistant crops (HRCs) designed to tolerate high levels of exposure to broad-spectrum herbicides, enabling farmers to spray only one heavy dosage per year, but this does not break the cycle of dependence upon chemical applications.

Genetically modified foods in U.S. markets include tomatoes, squash, yeast, corn, potatoes, canola, and soybeans (that are used in 60 percent of all processed foods, such as bread, pasta, candies, ice cream, pies, biscuits, margarine, meat products, and vegetarian meat and cheese substitutes). Genetically engineered foods not tested or labeled as genetically altered could jeopardize our health.

This process has already created some herbicide-resistant "super weeds," causing many farmers to have to spray even greater quantities of herbicides on their GM crops because the weed species have become even harder to control.

Cross-species transfers between fish and tomatoes or other unrelated species that would not have happened in nature might create new toxins, diseases, and weaknesses that can spread across species barriers.

This new combination of host genes and introduced genes has unpredictable

CASE STUDY: AVANT GARDENING

effects. These artificially induced characteristics can be passed on to subsequent generations and other related organisms. Transferring animal genes into plants also raises important ethical issues for vegetarians and religious groups.

Another form of genetic engineering is used to create "Bt crops" by inserting a genetically modified gene into a plant gene from a soil organism called Bacillus Thuringiensis (a pest-specific powder used, only when it is needed, by organic farmers and gardeners).

This inserted gene causes the plant to produce a substance that makes it toxic to certain insects (creating a built-in pesticide), and in theory, there should not be any need for chemical sprays. However, insects exposed to these transgenic crops over sustained periods of time might develop immunity to Bt, and even harsher pesticides will be needed to control the problem.

A more cosmic concern is raised by Harvard biologist E.O. Wilson in his book, *The Diversity of Life*. He estimates that we have identified and named only about 10 percent of the species that inhabit the earth. Knowing so little about our world, why are we in such a hurry to alter it?

Genetic engineering companies are carrying out a potentially dangerous global experiment by introducing large numbers of genetically engineered foods into agriculture and food supplies that might have unanticipated and harmful side effects leading to national or global food shortages.

More than 50 percent of the crops developed by biotech companies have been engineered to be resistant to herbicides. This could promote a rapid appearance of resistant insects, destroy the beneficial insects, or alter soil organisms and ecosystems. In addition, the pesticide produced by the plant might be harmful to the health of consumers.

There is no way of knowing the overall, long-term effects of genetically engineered foods on the health of those who eat them. Since most genetically modified foods are not labeled, manufacturers have already introduced genetically modified ingredients into many of our foods.

Labeling should be required for any food that contains a genetically engineered ingredient or has been produced using GM organisms or enzymes. This would help scientists trace the source of health problems arising from eating these foods. Food scares and epidemics are increasingly commonplace, and in response, the demand for organic food is skyrocketing.

Greenpeace has launched a new version of its popular Shopper's Guide, which is an online resource to help you find out whether the food in your shopping basket

CASE STUDY: AVANT GARDENING

is GM free. Hundreds of products are listed at: **www.greenpeace.org.uk/gm/gm-shoppers-guide.**

The importance of biodiversity includes sociocultural, economic, and environmental elements. Genetic biodiversity provides not only healthy crops, but it also allows for new plant and seed varieties, maintains soil fertility and its microorganisms, and makes soil and water conservation a priority.

Agricultural diversity maintains our bio-diverse plants, seeds, animal food sources, croplands, pastures, rangelands and the microbial and fungal sources necessary for healthy soil.

Another growing objection to genetic engineering is that we do not need to figure out how to grow more food. According to the Institute for Food Development Policy, nearly one-third of the world's land area is used for food production, and we already grow more than enough to feed everyone.

If the vitality, biodiversity, and health of our soil and crops can be improved, plants will be naturally resistant to pests and disease. We need to educate our farmers about the benefits of bio-diversity, soil sustainability, plant and animal health, natural pesticides, composting, and companion planting.

It has been estimated that only 1 percent of pesticides applied to crops reach the insects they are designed to kill; the other 99 percent pollutes the air, soil, food, and water, kills wildlife, and depletes the soil's vitality.

In the past, it was acceptable for farmers and gardeners to buy and spread chemicals and pesticides over their crops instead of understanding the mechanisms of sustainable organic growing methods and the importance of bio-diversity.

Maybe, as consumers demand more organic foods and growing methods, governments, agribusinesses, giant chemical companies, farmers, and home gardeners will be motivated to eliminate the use of genetically altered seeds and plants, carcinogenic pesticides, herbicides, and fungicides.

Maybe this will grow to include the livestock and fishery industries currently using antibiotics, chemical food additives, growth regulators, and hormones.

Courtesy Frank and Vicky Giannangelo, Avant Gardening, whose book, Growing with the Seasons, published by Sunstone Press, Santa Fe, NM, available in 2008.

Trash: That Other Toxic Waste

Some call the United States the "Consumer Nation." The United States is the number one trash-producing country in the world. Each individual generates 1,609 pounds in an average year. Translated into global numbers, this means that the United States, representing less than 5 percent of the world's population, generates 40 percent of the world's waste.

According to a 2000 report by the United Nations University's World Institute for Development Economics, there was about $125 trillion worldwide in household wealth, comprising everything from financial assets and debts to land, homes, and other tangible property. According to the International Monetary Fund (IMF), the United States holds 37 percent of the world's wealth. If we take this paragraph, and the one above, as statistical measurements, we can equate trash with wealth.

Waste as "Junk Mail"

Every time you check your mail, at least half the content will be items you glance at and throw away. Some are glossy, single-page flyers; others are letters, mailed in envelopes, from firms you have never heard of. The average landfill contains 37 percent paper. According to the EPA, this translates into the equivalent of 1.5 trees delivered to every household in the country each year as junk mail, or 100 million trees per year for the entire country. Only 22 percent of this is recycled. We pay, jointly, $370 million annually to get rid of this waste. Where does all this junk come from?

Every time you buy a product online, enter a sweepstakes for a product or promotion, join a club or organization, or use a credit card, your name goes on a list. Whether you buy a house, a car, a magazine subscription, or a gym membership, your name and address are being recorded. This information is added to a list in someone's database, and the list is later sold to direct marketers.

Direct marketing is a type of sales in which the agent sends his or her message directly to the consumer instead of via another media, such as television. The most common form of direct marketing is mail; the second is telemarketing. Both are equally annoying. The third, and quickly rising to the number one spot, is e-mail marketing, or spam.

Since most of us hate all three, why do companies continue to use these methods? Because, one time in a hundred, product information via direct marketing strikes a chord in the consumer. We see it advertised, feel the need, and respond. It takes only a 1 percent success ratio of contacts to sales to fuel the engine of direct marketing. Direct-marketing costs are low; companies pay a bulk rate on postage, install ten phone lines at commercial rates, hire people to work on commission, and pay for a single Internet hookup to send e-mails. A 1 percent sales-to-contacts ratio makes, or breaks, a start-up company.

You can get off these lists, but it takes time. You have to write to each of the solicitors in person and request to be taken off their list. Alternatively, you can visit the Direct Marketing Association Web site, and for one dollar, file a "blanket" request. Go to: **www.dmaconsumers.org/cgi/offmailinglist**.

If you do not have Internet access, you can mail your request to: Mail Preference Service, Direct Marketing Association, P.O. Box 643, Carmel, NY 10512. The form is available online but requires typing in a string of verification letters. It also requires you to enter your e-mail address, which made me, a consumer, suspicious. Would I be added to yet another, Internet junk-mail, list? If you want to try sending a letter, be sure to include your full name, complete address, and zip code.

Unfortunately, this will not remove you from all mailing lists. If the mailer does not subscribe to the Direct Marketing Association, or if the sender is a business you frequent, the mail continues. You also will continue to get unsolicited mail from local merchants, professional associations, political candidates, and that unfortunate mail addressed to "Occupant." Other than contacting these entities individually, you cannot prevent their exhortations to buy. Your Mail Preference Service application does not apply to business addresses, and the United States Postal Service will not withhold junk; anyone who can afford a stamp can send you mail.

You also can go to **optoutprescreen.com** and ask to be removed from these lists. This site asks for your birth date and your Social Security number, which is mildly disturbing. The information is not mandatory, but the form you fill out does have to be printed and mailed to confirm your request. Direct-marketing business is big money. It is no wonder the industry does not want you to find ways to disconnect.

Excessive Packaging

Remember the last time you bought a children's toy? There was a box, sometimes shrinkwrapped. Inside the box was a plastic covering, Styrofoam, or cardboard. The toy was attached to the packaging using plastic wires. Then there are pages of instructions in three or four languages. The plastic probably contains phthalates, a potent source of the xenoestrogens mentioned earlier. Manufacturers make more than 800 million pounds of phthalates each year, most of them packaging components that subsequently end up in landfills.

Just about everything you buy is packaged. According to manufacturers, this packaging protects your purchases from getting soiled or damaged and prevents product tampering. The packaging ends up in landfills.

Packaging accounts for a remarkable 90 percent of a product's cost. The other 10 percent is the cost of the product itself. That bottle of painkiller that costs you $5 costs about $.50 to make. You, the consumer, ultimately pay for the packaging, yet you have no say in it, and the United States government does not regulate against excessive packaging. In Europe, where cost- and environmentally conscious consumers have a say, the Liberal Democrats in England are preparing to demand restrictions on packaging. Like the recent Canadian law that requires full ingredient disclosure on cosmetics, this proposed packaging measure might make American manufacturers rethink their packaging strategies.

Up until now, packaging strategies and technologies have focused on what is called "cradle-to-grave" packaging, which guarantees a product whose waste must eventually be dealt with by the end-user, namely, you. But what if the plastic wrapper you throw in the garbage were engineered to melt in a few hours? What if that egg container were not only biodegradable, but also contained nutrients to renew and revivify the soil? This kind of packaging obviously would be more expensive, until technologies improve, but the cost savings in landfill disposal would offset part of that cost, and the rest could be accomplished via tax relief for products daring to undertake this "brave new world" sort of thinking.

This "cradle-to-cradle" concept, a collaborative effort from architect William McDonough and chemist Michael Braungart, is revealed in their new book *Cradle to Cradle: Remaking the Way We Make Things*. An overview, by author David Newcorn of *Packaging World*, contains excerpts from the book. To learn more about sustainable packaging, go to **www.packworld.com/view-16013** and **www.sustainablepackaging.org/about.htm.**

The Sustainable Packaging Coalition is a project of GreenBlue, whose focus is on redesigning industry by providing resources, solutions, and opportunities for sustainable practices. Green Blue's Web site is **www.greenblue.org**. The Sustainable Packaging Coalition provided the next Case Study:

CASE STUDY: ORIGINS & RENEWABLE ENERGY

Andrew King and Green Blue

1. Abstract

Origins Natural Resources, Inc. has a strong environmental ethic embedded in its mission. Origins has used a new product launch as an opportunity to push this ethic up the supply chain, enabling Origins to create a package consisting of 50 percent post-consumer recycled content paperboard and manufactured, printed, and folded using 100 percent renewable energy on a particular product line.

CASE STUDY: ORIGINS & RENEWABLE ENERGY

Through cooperation with suppliers whose product methods overlap with Origins' environmentally aware image, the corporation is able to offer a premium-quality recycled package made with renewable wind energy while remaining within a profitable budget.

2. The Players

Origins Natural Resources, Inc. Origins is a wholly owned subsidiary of The Estée Lauder Companies Inc. Origins has adopted an environmentally conscious position at a fundamental level in its mission statement, making it an attractive product for environmentally concerned consumers. Origins is currently producing a line of products in collaboration with integrative health specialist Dr. Andrew Weil.

Mohawk Paper Mills, located near Albany, New York, manufactures printing papers. Its use of wind power for 38 percent of its electric energy requirements demonstrations a commitment to environmental stewardship.[1] Mohawk Paper offers a wind energy production option by purchasing RECs from various energy suppliers including Sterling Planet and Community Wind Energy. Based on internal historic monitoring, Mohawk produces on average 74,000 tons of paper per 45 million kWh consumed. Using this ratio, the company calculates the tonnage as being paper produced using RECs it has purchased and can label a given tonnage as being produced with renewable energy.

This process is monitored by Green-e and the EPA Green Power Partnership. Mohawk offers papers made from virgin and recycled fibers certified by the FSC.[2]

In March 2006, Johnson Printing & Packaging, located in Minneapolis, Minnesota, arranged to have its entire production facility rely exclusively on wind energy. One of Origins' sister companies, The Aveda Corporation, is a key account with Johnson

- Pushing green practices up the supply chain.
- Renewable energy premiums offset by savings from energy conservation audit.

Printing & Packaging. With the backing of this powerful relationship, Aveda asked Johnson to explore wind energy options. This request afforded Johnson the opportunity to fully investigate wind energy and discovered the cost increases were low enough to warrant using wind energy exclusively to power all its entire

CASE STUDY: ORIGINS & RENEWABLE ENERGY

130,000-square-foot production facility. The broader effect is that Johnson now produces all of its packages with renewable energy and is positioned to serve a growing niche market of environmentally conscious consumers. Johnson purchases its wind energy credits from Xcel Energy, which typically charges an average surcharge of $2 per 100 kilowatt-hours.[3] Now Johnson's entire production energy bill is covered by RECs.[4] In compliance with the Minnesota Environmental Quality Board's recommendations, Johnson installed more energy efficient lighting saving $25-$50 dollars a day, virtually offsetting the increased cost of purchasing RECs.

3. The Story — Collaboration in the Supply Chain

Through its basic position promoting environmentally sound practices, Origins/Estée Lauder's Vice President of Packaging, John Delfausse, suggested assessing the opportunities of using renewable energy in packaging production while simultaneously continuing to use high-quality paperboard packaging with higher levels of post-consumer recycled (PCR) content. This project required finding a source of paperboard produced with wind energy and having it folded in a facility run on wind energy. Connecting these elements of the supply chain required consistent collaboration between Origins and its two suppliers for this product. After Origins failed to find a traditional packaging paperboard supplier willing to offer renewable energy options and adequate PCR content, Origins turned to Mohawk as a source for the paperboard. Through Aveda's established relationship with Johnson Printing & Publishing and corporation with Aveda, Origins was able to incorporate this new supplier. Origins' marketing staff then actively managed the cooperation between Mohawk and Johnson Printing & Packaging during the months of product trials.

- Flexibility when re-evaluating minimum specifications.
- Sourcing material from non-traditional suppliers

Origins worked with Mohawk Paper and Johnson Printing & Packaging on this initiative for approximately two years from initial concept development to product rollout.

4. Sourcing Paperboard

Originally, Origins wanted 18-point bright white paperboard packaging made of 100% post-consumer content using renewable wind energy. Origins thoroughly searched for

CASE STUDY: ORIGINS & RENEWABLE ENERGY

such a product with a range of packaging suppliers and was unsuccessful. Origins then approached Mohawk Paper, which actively promotes its wind energy and FSC certified PCR materials usage. Mohawk Paper specializes in commercial paper production for commercial printing purposes rather than for traditional packaging materials. Origins marketing professionals worked with Mohawk testing various layering combinations of thinner stock made with high levels of post-consumer content.

[1]"Mohawk and the Environment" Mohawk Web site. 25 July 2006
www.mohawkpaper.com/pdfs/5.%20Mohawk_and_the_Environment.pdf

[2]"Mohawk and the Environment" Mohawk Web site. 25 July 2006
www.mohawkpaper.com/pdfs/5.%20Mohawk_and_the_Environment.pdf

[3]"Xcel Energy's Windsource program a hit in Minnesota." Xcel Energy News Release. 6 August 2005. 10 July 2006.
www.xcelenergy.com/XLWEB/CDA/0,3080.1-1-1_15531_18513-19830-2_171_256-0.00.html

[4]"Johnson Printing & Packaging converts to Wind Power." Company News, 1 March 2006. 10 July 2006.
www.jppcorp.com

Waste from Plastics

Plastic comes from petroleum, a diminishing resource. This is a mixed blessing for those of us waiting for the post-petroleum world. Plastic is toxic. Polyvinyl chloride (PVC), otherwise called vinyl, is second only in production to polyethylene terephthalate, (PET). PVC is used primarily in building materials, such as window frames, doors, walls, paneling, water pipes, flooring, wallpaper, blinds, and shower curtains. It also is found in credit cards, records, toys, office folders, pens, and furniture. It is one of the world's largest dioxin sources. Dioxin is second only to radioactive waste in its toxicity, as the survivors of Love Canal (in Niagara Falls, New York) now realize. Polyvinyl chloride often contains phthalates, or plastic softeners, as found in baby toys. These phthalates are now being linked to a host of reproductive defects and cancer, as researchers make the link between toxic chemical buildup in fat cells and long-term exposure. Before 1991, the waste products of manufacturing polyvinyl chloride, notably ethylene dichloride tars, were burned on ocean-going incinerators. When that practice was banned, manufacturers resorted to burning these tar wastes in incinerators or dumping them into deep wells, where eventual leaching into groundwater will return them to our environments.

PET is used to make soft-drink bottles and other food containers. Bottled water is now a $2 billion-per-year industry. Contrary to popular belief, the FDA does not require an expiration date for bottled water, though many bottlers do post a date on the product label, usually two years after production.

Dr. William Shotyk, director of the Institute of Environmental Geochemistry at the University of Heidelberg in Germany, researched the effects of storing water in plastic bottles made from PET and found traces of antimony — a chemical used in making PET — in almost all bottles sampled, with concentrations of antimony doubling in bottles that had been stored for three months or more.

The federally established limit for antimony is 0.0002 milligrams per cubic meter (mg/m3) inhaled; there are no established limits for ingestion. In animal studies, inhaled antimony produced lung tumors, cardiovascular effects, and kidney damage. In humans, long-term exposure can result in loss of pulmonary function, chronic bronchitis, chronic emphysema, pleural adhesions, and cardiovascular difficulties. Antimony also has been linked to spontaneous abortion in humans. Taken internally, antimony was shown to significantly affect the blood, liver, nervous system, and intestinal tract of laboratory animals. Ingesting small doses of antimony will cause mild stomach upset and depression. In larger quantities, antimony causes cramps, projectile vomiting, and death.

The quantities found in the water bottles were well below officially sanctioned limits. Unfortunately, these limits have not been adjusted since 1991, the same year in which recycling programs became common because of an overabundance of plastics in landfills. If you must buy plastic, which has become ubiquitous in the modern world, remember these safety tips:

- Do not microwave food or beverages in plastic or Styrofoam containers. Use a paper plate, a paper napkin, or glass or ceramic dishes.

- Use metal or wooden cooking utensils instead of plastic.
- Store foods in glass or ceramic jars rather than plastic containers when possible.
- Select food and beverage products that come in glass, rather than plastic, containers to discourage manufacturers from continuing their use of plastic.
- Do not buy soft "chew" toys for your baby, pacifiers and teething rings, or bottles and nipples without first checking the ingredients used in their manufacture. When in doubt, go online and find an environmentally friendly supplier.
- Do not buy plastic lawn furniture, imitation Naugahyde furniture, or other plastic furniture that your children or pets may chew on.

A United States Center for Disease Control and Prevention study of 2,400 individuals completed in 2002 found, in addition to the expected chemicals such as lead and second-hand smoke, 24 other compounds that had never been observed or measured in the human body, most of which did not exist in the 1930s. These compounds included 76 chemicals linked to cancer. The participants had a total of 48 polychlorinated biphenyls (PCBs) first made by Monsanto and banned in 1976. Unfortunately, they are still used in other countries, and they persist in the environment for decades. You can find this report at **www.cdc.gov/exposurereport**.

A study conducted at Midway Island in 1996 showed PCBs occurring in albatross populations in "concentrations that were significantly greater than the method detection limits," or higher than the scientific apparatus had been calibrated to measure.

Recycling

Reports differ as to whether our nation is learning to recycle. According to Neil Seldman, president of the nonprofit Institute for Local Self-Reliance (ILSR) in Washington, the outlook is promising. In an interview with USINFO, Seldman speculated that 80 percent of construction and demolition waste is currently being reused, compared to 55 percent five years ago.

In terms of recycling, California leads the pack nationally, with San Francisco posting a ban on plastic bags effective in the spring of 2008 and Oakland following with a similar ban, effective in the same time period. The ban applies to stores selling $1 million worth of goods per year, which would include most supermarkets, chain drugstores, and other large retailers. These stores would be required to offer shoppers paper bags instead, and these bags would be completely recyclable and contain at least 40 percent recycled material. In July of 2007, a California mandate forced retailers to provide on-site plastic bag recycling facilities. Los Angeles was the first city in the country to declare a zero waste goal. L.A. is currently at 62 percent and aims for a 90 percent reduction by 2025.

The United States makes about 3 trillion bags annually, a figure plastic manufacturers dispute, but the evidence lies in our landfills. Their manufacture consumes 13 million gallons of oil. Each bag costs the retailer about $.05-$.10, and the cost is passed on to the consumer through higher retail costs on everything from a bottle of shampoo to a bunch of bananas. Currently, there is a Texas-sized "island" of trash, composed largely of plastic bags, floating in the North Pacific called The Great Pacific Garbage Patch, or Eastern Garbage Patch, all courtesy of our wastefulness. The oil used in the manufacture of these bags is enough to heat 130,000 homes.

Some firms, such as Aldi's supermarket, supply or sell bags that can be reused. Others, such as Cub supermarket, sell large cardboard boxes for the same purpose. Some environmentally conscious firms, such as Peets, Starbucks, Wild Oats, Whole Foods, and Trader Joe's, are offering financial incentives to customers who supply their own bags. Ikea charges for bags, hoping to motivate customers to recycle and reuse.

Another major problem is plastic bottles, most of which are made of PET. This substance degrades slowly in landfills, leaving toxic residue. A 2006 report on bottle recycling, conducted by the As You Sow (AYS) Foundation and the Container Recycling Institute, found that 10 out of 12 major (PET) bottle manufacturers scored a D or below on their recycling efforts. PepsiCo got a C, or 2.3 out of 4, by using 10 percent recycled material in 2005. Coca-Cola also got a C, or 2.1, but lost out to Pepsi because it has not committed to a 2006 goal of using 10 percent recycled plastic in its containers. It did, however, score best in reducing extraneous packing and materials. All the rest, including Miller and Molson Coors — which merged in October 2007 to become MillerCoors — Polar, Starbucks, and Nestle Waters, failed. Cadbury Schweppes, Cott, and National Beverage did not even reply to the survey. You can find the report at: **www.container-recycling.org/publications/reports/scorecard.htm**.

In spite of the numbers of plastic bottles sold every year, bottle recyclers and recycled plastic users are reporting a shortage of materials. The finger clearly points to us, the consumers. Our landfills are filling faster than we can locate them. One hundred dumps closed in 2006, filled to capacity. By 2057 all the existing landfills will reach capacity.

As for what is in these landfills, there is no scientific study, but operators were asked to observe and estimate the contents of their landfills. The primary component is always paper. According to an article on the Energy Information Administration's Kids' Page, one waste archaeologist reported finding a newspaper from the 1970s that was still fully readable. Paper accounts for 31 percent of a landfill's volume, containers and packaging, another 27 percent. Yard wastes are 10 percent, and things such as furniture, tires, and appliances contribute 32 percent. No one knows what is going into landfills and how quickly or slowly it degrades, since landfills are monitored for leakage and illegal dumping but not content.

Aluminum, used in soda cans, does not biodegrade, can be recycled endlessly, and represents the safest method of containing beverages. According to the Aluminum Can Association, recycling of these cans is at more than 51 percent. Figures from the EPA show recycling efforts recapture only 44 percent, and slightly more than one-third of these throwaway cans were rescued from landfills after people had disposed of them.

Recycling How-Tos

Most cities and municipalities in the United States offer some form of recycling program that the consumer can use to recycle paper, plastic, bottles, and cans. Budget crunches and inflation have reduced some of these programs to skeletal proportions, — no curbside service and recycling facilities open only one day per week — but you can still participate in weekly recycling for household waste and annual or semiannual recycling events for such things as Christmas trees, electronics, oil, and batteries. My city, for example, has a one-day-per-week recycling program and a two-sort program for cans, plastic, and glass or paper products. Wide-neck plastic containers, such as peanut butter jars, cannot be recycled, nor can greasy pizza boxes. Electronics recycling also is prohibited, except through the Waste Management division. In my state, putting a television or a computer monitor in the trash is illegal, but there are local electronics stores that will take my used equipment. I can buy a special recycling bin for $5.50 plus sales tax and use it to dispose of my sorted household waste.

If you do not have curbside pickup for recycling items, you can contact your local or state waste-control program and ask how that agency wants recyclables sorted. Then all you need do is buy another trash container, sort the waste into garbage bags, and haul it to a recycling facility. If you have a fireplace or wood stove, you can burn that useless paper; just be sure to use chimney cleaner regularly, as burned paper produces an inordinate amount of creosote, which plugs chimneys and causes fires. Cans and bottles can be crushed in a counter-mounted device that costs about $10, paper and cardboard can be flattened into bundles.

Batteries can be recycled through the American Automobile Association or through your local automobile repair shop. Oil can be recycled the same way. Hazardous wastes, such as old cans of oil-based paint or lacquer thinner, have to be taken to an approved landfill. Do not dump them down the drain; they go directly into the water supply. Lacquer thinner in particular is classified as an F1 hazardous waste by the federal government. For more information on recycling how-tos, go to **earth911.org or protectingwater.com/recycle.html**.

Last, join the Freecycle Network, a grass-roots, global organization of 4 million members who are giving, and getting, free stuff for themselves and their communities. Started in 2003 by Deron Beal, the organization is devoted to sustainability, as people from all walks of life band together to turn "one man's trash into another man's treasure." Check out the group's site at **www.freecycle.org**. Membership is free, and you will be doing your part to reduce your carbon footprint.

Remodeling

This section will look at the more expensive ways to make a house environmentally friendly through remodeling. Included are such items as creating a residential "green" roof, sealing or replacing windows, green alternatives to commercial carpeting and other flooring materials, new eco-friendly paints and stains, low-flow and other eco-friendly toilets, efficient lighting and energy systems, and using salvage facilities to recycle and save money on your remodeling project.

Roofs: Not All "Green" Roofs are Green

First, I want to clear up a misconception. "Green" roofs are not the same as environmentally friendly roofs, which use special, eco-friendly roofing materials. I will cover both, but it is important to understand the distinction in advance. A green, or living, roof is partially or entirely covered with growing plants. An environmentally friendly roof is built from energy-efficient, reflective substances designed to direct light and heat away from the building and renewable or sustainable building materials, such as recycled wood, with waterproofing materials made from corn.

Green Roofs for City Dwellers

It is now very stylish to have a "green" roof. Having one marks you as someone who is aware of the environmental problems we face in this, the first decade of a new century. A green roof says you care. More important, it says you care not just about today, but about the world of tomorrow, in which future generations will live.

The average city rooftop is layered with black tar, which traps sunlight and heat, raising the temperature of the surrounding air and the temperature inside the building. If your home is a co-op apartment, you can get together with your neighbors or building association and put in a green roof with surprisingly little investment on each person's part. Green roofs in condos or apartments:

- Provide a green and growing retreat for city dwellers
- Increase the life of the roof
- Reduce storm-water runoff and consequent waste
- Provide excellent sound insulation from noisy streets
- Filter pollutants and carbon dioxide out of the air
- Provide locally grown food
- Increase wildlife habitat in metropolitan areas
- Reduce heating and cooling loads on a building
- Reduce the urban heat-island effect

Green roofs offer so many benefits, both to the occupants of a building and to our shared biosphere, it is a wonder there are not more of them. Unfortunately, city planners and bureaucrats are slow to adapt to new ideas, so the progression from asphalt to living roof often is delayed by red tape and unfounded concerns, many generated by the straw-bale roof movement of the 1970s, which was aesthetically horrifying and a mess to clean up.

The major problem with a green roof, aside from improper installation, is its initial cost. Green roofs are expensive, but so are homes. After the roof is installed, maintenance costs are minimal, and green roofs last significantly longer than standard roofs. The Hanging Gardens of Babylon endured for 1,000 years. More recently, the green-roof installation on the Rockefeller Center in New York, built in 1930, survives intact.

The last hurdle to making green roofs economically viable is convincing city governments to change their building codes and provide incentives to property owners who want to go green from the top down. So far, the city of Chicago leads the change, providing $5,000 to people willing to invest in green-roof projects. Other notable projects include the top of the Ford Motor Company's plant in Dearborn, Michigan, and San Francisco's living roof atop the California Academy of Science building, which was five years in the making and contains 1.2 million plants.

For more information on green roofs, go to **truths.treehugger.com** or **www.greenexchange.com.**

Green Roofs for Homeowners

Homes cannot always be retrofitted for some green-roof applications, such as shrubs and trees. For one thing, most residential homes have a peaked, or sloped, roof. The pitch of this slope is dependent primarily on snow loads. In areas where snow falls deeply, the roof's pitch must be greater than in areas where snow is an occasional occurrence. Roof pitches are measured like stairs, in rise and run. Most homes have 6:12, the 6 representing the rise of the roof and the 12 representing the run, or length. For example, if your home is 24 feet wide, half the roof is 12 feet. In that 12 feet, the roof rises 6 inches for every foot, or 72 inches to the peak.

A 6:12 roof is common in snowy areas, such as the Midwest. In areas of the Sierra Nevada, where snow falls six feet deep at a time, the pitch might be closer to 7:12 to permit snow to slide off before it collapses the structure. Modern roof trusses have eliminated some of this danger, but different geographical areas of the country will have pitches mandated either by the local or municipal governments, and you must know what these are before you repair or replace a roof or if you are building a new house. In many areas, this pitch prevents some of the more spectacular green-roof applications: you cannot walk on that kind of grade, and you cannot effectively put plants, even plants in pots, on a slope that steep. You can, however, still have a green roof. After all, green roofs have been around since the Middle Ages. Back then, they were called "soddies." Irish immigrants brought the technology to the New World, and many settlers covered their roofs with sod, both to keep their houses warm and because sod, cut from the virgin prairie, was cheap, plentiful, and easy to maintain.

A green roof sounds complicated but is quite simple. The basic elements include your roof's underlayment, a waterproof membrane, a drainage element, a layer of soil, and vegetation. Using your existing roof, sans shingles or tiles but including the plywood base, you can attach boards to the fascia in a vertical position and equipped with drain holes to act as a soil dam, then put down a waterproof layer and install a drainage layer. Draincore2 is a good choice. After that, you also can add perforated pipes at intervals to affect more drainage. Finally, put down a layer of soil and spread sod. Voila — you have a green roof. Green roofs replace an impervious surface with a heat-absorbent surface, which cuts heating and cooling costs. They are environmentally sound, because grass, like every other green plant, absorbs carbon dioxide and provides oxygen. They also are pretty. You can choose plants other than grass, or intersperse the grass with other plants, but plants will take longer to root and form colonies. You also must consider your regional climate when planting a green roof; grass that does wonderfully in Iowa with a little maintenance will not do nearly as well in Arizona.

Plants for Residential Green Roofs

Residential green roofs, in spite of their environmental IQ, do not offer infinite opportunities for creativity within the plant kingdom. Residential green roofs are hot, drier than the surrounding landscape, exposed to wind, and prone to greater temperature extremes than your yard. Plan your roof as though you were xeriscaping; water runs downhill, so the soil closest to the fascia will always be wettest, and plants at the peak will be buffeted by the wind.

Plants that grow wild in your geographic area are a good choice, as long as they do not have extensive root systems that might penetrate the waterproof layer. If you live in an area that has very cold winters, some plants might not be sufficiently frost-hardy. Roadside planting of native species has become common, and your state transportation agency may have a list of hardy, native plants. Achillea grows almost everywhere and is tolerant of extremes of temperature and water. Asters, coreopsis, Ajuga, or bugleweed, black- or brown-eyed Susans, coneflowers, flowering mint species, forget-me-nots, wild indigo, lupine, rue, and flowering sage are also good choices. You might want to choose from a group of plants known as alpines, which grow in harsh conditions in mountains from Europe to the United States. Backyard Gardener (**backyardgardener.com**) is an excellent resource and offers alpines from seed as well as a list of regional alpine growers. When possible, choose low-profile plants to keep them from being damaged by wind.

Composition Roofs

Asphalt shingles come in two different kinds: organic-based or fiberglass-based. Organic shingles are made from materials such as recycled waste paper, wood fibers, and felt, then saturated with asphalt coating and topped with colored granules. Organic-based asphalt shingles contain about 40 percent more asphalt per 100 square feet, which gives them more weight, durability, and wind resistance. In spite of their "organic" label, these shingles, with their inordinate burden of asphalt and chemicals used in recycling the ingredients, are not truly environmentally friendly, though they do represent sustainability. This is equally true for many recycled products, and I make the distinction only to point out the essential difference between environmentally friendly and sustainable; the first is good for us and for the earth, and the second is good for the earth but not always for us.

Fiberglass-based shingles are made of a layer of fiberglass under a layer of asphalt, topped with a layer of colored granules. Fiberglass is a composite of glassine fibers spun from melted glass. During the manufacturing process, certain chemicals are employed, among them polycyclic aromatic hydrocarbons, arsenic, coal-tar, bitumen, quartz, chromium, formaldehyde, and strong acids. Polycyclic aromatic hydrocarbons are one of the most widespread organic pollutants. Some are known carcinogens. To increase reflectivity, the shingles are coated with a reflective substance, such as titanium dioxide or silicone. Material Safety Data Sheets list titanium dioxide as a cancer hazard. Formaldehyde, associated with the Aspartame scandal, is known to cause immune system and nervous system changes and damage, as well as headaches, poor health, irreversible genetic damage, and a number of other serious health problems. Asphalt shingles of either kind have a tendency to emit this formaldehyde into the air on hot, sunny days, an effect known as off-gassing, which contributes to air pollution.

If you currently have an asphalt shingle roof and are considering an upgrade to more environmentally friendly products, you also should be aware that every year about 11 million tons of asphalt shingles make their way into landfills. The technology to recycle them is in its infancy.

Recycled-Product Roofs: What to Avoid

Recycled rubber tiles are all the rage. People assume that rubber, a natural compound, is organically safe. Current research at Bucknell University indicates that leachate from tires discarded at landfill sites can kill entire aquatic communities, from algae to fish. At lower concentrations, these leachates can cause reproductive problems and precancerous lesions.

Part of the toxic nature of rubber is due to its mineral content, which includes aluminum, cadmium, chromium, copper, iron, magnesium, manganese, molybdenum, selenium, sulfur, and zinc, all of which have been identified in field leachate studies. Zinc, which comprises as much as 2 percent of a tire's mass, is a particularly potent biological agent that can kill plants. Rubber leachates also contain plasticizers, such as polyaromatic hydrocarbons and accelerators, such as 2-mercaptobenzothiazole, which is used during the vulcanizing process. These are toxic compounds whose persistence in the environment already has been noted and whose effects can include skin and eye irritation, neurological damage, organ damage, and even death.

Enviroshake, a new composite product by Wellington Polymer Technology Inc., in Chatham, Ontario, is a simulated cedar shake made of recycled materials reclaimed from manufacture and building demolition. It is composed of plastic, rubber elastomers from tires, and cellulose fibers. Enviroshake is durable, provides protection from ultraviolet light, is wind and hail resistant, and resists mold, mildew, and insects. It does not require pre-treatment, is easily installed, and is maintenance free.

There are some panel roofing systems that duplicate the above ingredients and come pre-formed to resemble cedar shakes. Forest Products Laboratory has one such panelized system. Authentic Roof is another manufacturer of recycled roofing products. Its roofing products are made of thermoplastic polyolefin (TPO), a material made by combining polyethylene, polypropylene, and ethylene propylene diene monomer (EPDM) rubber. Thermoplastic polyolefin is an alloy, not a composite; that is, the materials are bonded at a molecular level and cannot be separated, so degradation, outgassing, and leaching are less likely to occur.

Though many recycled roofing products have earned the Energy Star label, roofing products made from recycled wood also contain varying amounts of rubber, plastic, and other water-retardant chemicals whose outgassing or leachate products contribute to pollution and thus to global warming. Although recycling itself reduces one's carbon footprint, recycled products do not have enough history to verify their benefits or establish their dangers. Because they are so new and essentially experimental, costs also are quite high.

If you can afford it, choose one of the next roofing products.

Metal Roofs

Metal roofs are not the eyesore they once were. Metal can be molded into shapes reminiscent of slate tiles or shingles, and metal roofs now come in many colors. Metal roofs do not burn and help prevent fire from spreading to your home during a forest fire. Metal roofs last a long time, require little or no maintenance, and resist decay, discoloration, and mildew. Modern metal roofs are highly wind resistant, shed rain and snow equally well, and are generally resistant to hail. They also are thermally efficient, and good ones provide a natural radiant barrier effect that reflects 97 percent of radiant heat back into the atmosphere instead of into your home. Metal roofs can last 40 years, 10 years longer than cedar shingles. Ongoing costs for maintenance are about $0.57, compared to $3.57 for asphalt-shingle roofs.

The single drawback to a metal roof is the coating, which is commonly polyvinylidene difluoride (PVDF). PVDF is a highly nonreactive, pure thermoplastic fluoropolymer. It is also known as Kynar or Hylar. Although no health or environmental hazards are reported during normal handling, if this material is heated in excess of 600°F, the average, initial temperature inside a burning house, hazardous by-products will be emitted, including hydrogen fluoride and oxides of carbon. Hydrogen fluoride can be fatal if absorbed through the skin, swallowed, or, as is most likely in the case of a house fire, inhaled. Exposed skin may be burned, and hypocalcemia, or calcium depletion, may occur. A victim may experience pulmonary edema and obstruction of the upper airway due to swelling.

Environmentally Friendly Roofing Materials

If you are replacing your roof or building a new home and focusing on environmentally friendly alternatives, choose clay, concrete, slate tiles or wood shingles.

Clay or concrete tiles are made from earth-friendly, natural products; clay is a form of earth, and concrete is cement, which incorporates calcium in the form of limestone, aggregate, and water. Concrete and clay tiles are extremely durable, able to withstand wind, hail, rain, earthquake, and fire, a valuable characteristic for homeowners in California. They are energy efficient, reducing heating and air-conditioning costs by 40 percent. They emit no toxic fumes. They require much less maintenance than other roofing products. Sealing is not necessary but is advised. Commercial sealants, like asphalt shingles, emit potentially toxic volatile organic compounds into the air.

Authentic slate tiles, not imitations made from recycled rubber and plastic, are another environmentally friendly choice. The use of slate for roofing dates to the American Colonies. Slate was locally available along the Eastern Seaboard but difficult to transport in the 1600s, so poorer settlers used wood shakes. By the middle of the 19th century, canals and railroads made slate more portable and less expensive. Its use declined with the advent of asphalt shingles in 1903. Slate is still expensive, $350 for 110 square feet compared to $50 for asphalt shingles, but its lifetime cost, including maintenance, is $2.14 versus $3.57 for asphalt shingles. Slate roofs last about 200 years, asphalt shingle roofs, a mere 20.

For those who simply cannot afford clay or slate, wood shingles are an inexpensive, earth-friendly alternative. Wood shingles are commonly made of cedar or other dense, non-rotting woods. Cedar does not have to be treated and will age gracefully. It is almost as strong as oak and has natural antibacterial and antifungal properties, making it resistant to rot and insects. Cedar also is easy to work with. Costs are about $300 per

100 square feet, and maintenance is three-quarters that of asphalt shingles. Cedar shingles last twice as long as asphalt shingles.

For more information, go to the following sites:

- **www.greenproducts.net/products/products.html**
- **www.sacredplaces.org/PSP-InfoClearingHouse/articles/Slate%20vs%20Asphalt%20Comparing%20the%20Alternatives.htm**
- **www.historichomeworks.com/hhw/pbriefs/pb04.htm**
- **www.greenchemex.org/about**

Windows

Homes built before 1975, the last historic "energy crunch," have little or no insulation due to a lack of adequate federal standards. Homes built between 1975 and 1985 have about half the amount of insulation required by today's standards. These homes lose an appreciable amount of heating and cooling, resulting in increased electricity and gas costs and consumption and inadequate performance of heating and cooling units.

Windows make up between 10 and 25 percent of a home's exterior wall space and account for 25 to 50 percent of heating and cooling costs. Windows gain and lose heat through a process known as heat transference, an elemental principle of thermodynamics. If you put a spoon in a cup of hot coffee, the spoon gets hot. A window heated by the sun makes the room warmer. Thermodynamics treats heat and cold not as substances, but as energy that can pass through various mediums at a molecular level. By the same principle, it is not the electric line itself that powers your microwave but the energy passing through the line.

In winter, your windows are transferring, or conveying, heat via the glass into the outside but also conveying cold air inside, because cold air "weighs" more than hot air, or has greater molecular density. This convection is particularly noticeable in older, wood-framed, single-pane windows. Older, double-pane glass windows address some of this problem by providing a barrier through which heat or cold must pass. These windows provide an air space through which outside air must pass, and the adjusted temperature of the air in this space, mediated by your home's temperature, slows convection to some degree. Modern windows provide this air space but fill it with a non-conductive gas such as argon, which halts convection almost completely. Andersen Windows, one of the largest window manufacturers in the country, uses argon in its insulated glass windows. Other gases being used are carbon dioxide, sulfur hexafluoride, or SF6, and argon-krypton mixtures. Skyline Windows, a manufacturer of highly energy-efficient windows, also makes custom windows, including simulated leaded glass to match the exteriors of classic old homes, like the English Tudor home shown in the next Case Study.

CASE STUDY: SKYLINE WINDOWS

Courtesy Dean Talbott,

Minnesota Power, serving 141,000 electric customers in northeastern Minnesota and northern Wisconsin.

Architect: Cornerstone

Project Overview

The homeowner of an exclusive Greenwich, Connecticut, estate wanted to replace the old, decayed steel windows of an English Tudor mansion with modern, efficient windows that replicated the old design but were less expensive than new steel. The heavy sightline of wood casement products so common to suburban frame construction was unacceptable to him. An added challenge would be to re-create the thin lines of lead caming that would lend charm and authenticity to the home.

The solution involved selecting the same structurally glazed, open-out casement used so successfully in Skyline's renovation of an art deco apartment housed in New York City. The 2- inch perimeter frame with beveled "putty" bead profile would give the same delicate shape as the former steel while at the same time allowing the customer to enjoy the high performance of the aluminum thermal section. To complete the

CASE STUDY: SKYLINE WINDOWS

design, the glass supplier was asked to apply a lead caming tape to the number one and two surfaces of the insulating glass so that the "leaded look" could be achieved, while the interior glass became the protective insulating barrier.

Contacts

To learn more about this project, contact Skyline today at 212-491-3000 or e-mail info@skylinewindows.com.

The most important advance in window manufacture, after the development of double-pane, gas-filled glass, is the development of low-emissivity, or low-e, glass coatings. These coatings of silver or tin oxide permit light to pass through but reduce heat conductance, or U-value, and increase glass resistance to heat flow, or R-value. These coatings also prevent about 78 percent of harmful ultraviolet rays that can fade your drapes and furniture upholstery.

If you can afford it, your first remodeling project to reduce your carbon footprint should be new, high-performance windows. These windows should comply with the CAN/CSA-A440-M90 standards; these are the most thermally efficient windows you can buy. These windows offer the latest technology in gas fills, low-e coatings, framing materials, and low-conductivity spacers. Minnesota Power's Millennium Star Project is an excellent example of window efficiency as well as overall building-envelope efficiency and is provided next. After you read it, you can go on to investigate each of the aspects of energy-efficient windows, including gas fills, coatings, spacers, and frames.

MILLENIUM STAR PROJECT

The Project

Minnesota Power's Millennium Star project gives you a firsthand look at the construction of a high-efficiency home that can be heated and cooled for $300 or less per year — without sacrificing big windows, unique floor plans, fresh air, and all of the other comforts of modern living. If you are planning to build or remodel a home and you have super energy efficiency in mind, follow this project.

MILLENIUM STAR PROJECT

Our goal is to use home building technologies that are most efficient from an energy and resource standpoint but that remain in reach of a middle-class budget. The building products and techniques are monitored for heat loss and energy-efficient performance.

Minnesota Power's Millennium Star project evolved from a 1998-1999 Conservation Improvement Program (CIP) project. The state of Minnesota requires investor-owned utilities such as Minnesota Power to provide a wide range of energy-saving information to customers.

Project partners include LHB Engineers and Architects, Arrowhead Builders Association, and Environmental Pollution Authority Energy Star Home.

The Foundation

Historically, basements were perceived to be cold spaces in the winter months and damp, musty spaces in the summertime. Often, the basement level was used only for storage, mechanical equipment, and laundry facilities. With new construction costs averaging $100 per square foot in today's home market, it is a very cost-effective investment to construct a basement to be a dry, warm, and comfortable living space.

One of the key features of the building site for the Millennium Star Home was the gradual south-facing slope. This allowed for the design and construction of a walk-out basement (or a lower level, as we like to call it). The lower level is fully exposed to the sun with large amounts of south-facing glass, providing a bright and well-lit living space.

For the Millennium Star Home, we used the Arxx Insulated Concrete forms, manufactured by AAB Building Systems. The Arxx wall system consists of forms that are designed to remain in place after the concrete has been poured. The forms are made from expanded polystyrene foam and held in place for the placement of concrete by polypropylene ties called webs. The forms are laid up in courses, much like masonry blocks, with reinforcing steel inserted as per engineering requirements, via a snap-in system. The insulated concrete form has an R-value of 22, compared to only an R-2 for an uninsulated block or poured wall.

Each block is 4 feet long, 16 inches high, and 12½ inches thick (about the size of six standard concrete blocks but weighing only 6 pounds). The concrete wall thickness is 8 inches, and the insulated forms are linked together by polypropylene (plastic) webs. These webs are placed 8 inches apart and extend from the outer to the inner surface of the form. Sheetrock, siding, and finished wallboards are attached directly to the surface with screws or appropriate fasteners. Reinforcing steel is placed horizontally on each course and vertically every two feet. The polypropylene webs are grooved for the placement of the steel reinforcing rods.

MILLENIUM STAR PROJECT

After the walls are erected, window and door openings are appropriately placed, and the entire wall is secured and braced with the Arxx alignment system. The concrete is pumped or poured into the walls at a specified rate, and truing or plumbing of the wall takes placed immediately after the wall has been poured.

An adhesive-backed waterproofing membrane is attached vertically to the exterior of the wall on the below-grade surfaces. The soils at the building site for the Millennium Star home consist mainly of heavy, nonporous material, such as clay or silt. A free-draining or porous type backfill is used to drain water away from the wall and maintain a drier, more comfortable basement. Insulated drainage boards are also available that direct water to the footing while providing thermal wall insulation. An impermeable top layer is added to prevent ground adjacent to the foundation from getting saturated. Soil should be sloped away from the foundation to direct all surface water away from the building.

Combining the well-insulated R-22 basement wall, the drainage and waterproofing material around the foundation, the large amount of window area to the south, the two-inch rigid foam under the basement slab, and the hot-water radiant floor heat, you end up with an entire lower level that is a bright, warm, dry, and comfortable living space.

Building the Shell

An objective of the Millennium Star home is to incorporate many different types of building materials into the construction so that Minnesota Power can demonstrate and monitor the energy efficiency of the various products. Advanced infrared thermography, along with data loggers and sensors, will be used to determine the performance at the various thermal components.

The building shell of the Millennium Star home uses three different methods of wall construction:

- Structural insulated panels
- Staggered stud construction
- Advanced framing using various types of insulation

Structural Insulated Panels (SIPs)

Structural insulated panels (SIPs) are made by joining high-performance, rigid foam insulation to oriented strand board. The result is a building product that is predictable, resource efficient, and cost effective. SIPs are used as floors, walls, and roofs on all types of buildings.

MILLENIUM STAR PROJECT

Several of the walls in the Millennium Star home are constructed with 4-by-8-foot foam core panels that are 6 inches thick. The panels lock together, giving the walls an R-value of 23 compared to an R-value of 8 in more traditional wall construction. SIP buildings are more energy efficient, stronger, quieter, and more draft free than older technology systems, such as stud framing with common fiberglass insulation. And, 10-inch foam panels were also used for the highest part of the roof. It is called a "hot roof," as it is unventilated.

Window openings can easily be cut out of the panels, and wiring is fed through conduits that have been pre-drilled through the form insulation.

Staggered Stud Construction

The walls on the west end for the Millennium Star home are made with double 2-by-4s that become an 8-inch wall. Staggered stud construction eliminates the thermal bridging of wall studs and allows space for a high-density blown cellulose insulation, giving the walls an R-value of 30. Wall studs are placed at 24 inches on center with a single top plate. The roof trusses are lined up directly over the wall studs.

Advanced Framing

This type of wall construction will use less lumber, as the framing is 2-by-6-inch boards with studs at 24 inches on center with a single top plate. The roof trusses are lined up directly over the wall stud. Normal studs are at 16 inches on center.

Minnesota Power will be monitoring the performance of six different types of wall insulation in the Millennium Star home.

- Dense blown cellulose
- Blown-in fiberglass blankets
- Spray foam
- Standard fiberglass batts
- High-density fiberglass batts
- Structural insulated panels

Windows & Doors

Shopping for windows and doors can be a confusing experience if you are unfamiliar with window technology. Whether you are retrofitting an existing home or building a new one, selecting windows and doors is crucial in improving your home's energy efficiency. Other benefits of high-performance windows and doors

MILLENIUM STAR PROJECT

include increased comfort, reduced maintenance and repair bills, and lower energy costs.

Windows and doors have traditionally been a weak link in the home insulation system, but rapid improvements in insulating quality are being achieved.

Selecting Windows and Doors for the Millennium Star Home

We considered these key elements:

- Insulating quality of materials
- Air tightness of the window sash and frame
- Installation techniques that eliminate air leaks between the window frame and wall

For the Millennium Star Home, we wanted windows with an overall U-value of .35 or less.

We chose a Platinum II fiberglass frame window with three panes of glass manufactured by Accurate Dorwin in Winnipeg, Manitoba, Canada. There are two argon-filled air spaces with two low-e coatings and triple weather stripping. The spacer between the glass is non-metallic and made from extruded silicon, reducing thermal bridging at the glass edge.

R- and U-values are measurements used in rating how well a material resists heat conduction. The higher the R-value, the better the insulating quality. In the past, window performance was measured by center of glass R-value, which did not accurately reflect the total window performance. The more effective method is to compare the overall unit U-value, which includes the performance of all window components, such as frame, glass, and spacer. The lower the U-value, the better the insulating quality.

If your windows are 40 to 50 years old and have aluminum frames, double glass, and one-half inch air space, they probably have a U-value of about .62. If you are installing new windows, select windows with a U-value of .35 or less, as we did for the Millennium Star Home.

Lower U-value windows not only help lower energy bills, they also maintain higher relative humidity levels in your home without a condensation problem. Comfortable humidity levels are about 40 to 60 percent. Windows with a low U-value have a glass surface temperature of about 58 to 20 degrees warmer than standard older windows. The cooler the window surface, the greater the chance for condensation. Be sure to

MILLENIUM STAR PROJECT

find out whether the U-value rating for a window applies to the entire unit (including frame) or to the center of the glass only.

Here are some other energy-efficient options we considered in selecting the windows:

Multiple glazing: Increasing the number of glass panes (or glazings) improves the insulating quality of the window.

Gas-filled windows: The air space between the glass panes adds to the window's insulating quality. The R-value of the window increases even more if the space between the panes is filled with argon or another gas with insulating properties superior to ordinary air.

Low-e glass: Materials such as standard window glass that readily radiate heat to a colder object are said to have high emissivity. Ultra-thin layers of special metallic compounds, deposited on window glass, reduce radiant heat transfer and save energy. They are commonly called low-emissivity coatings.

Sash and frames: The sash and frame comprise only 10 to 20 percent of the total window area. Steel and aluminum frames are durable, but they have little resistance to heat movement. If you choose a window with a metal frame, it must have a thermal break to combat heat loss. (A thermal break is a section of material — usually polyurethane or wood — that is sandwiched between the frame parts.)

Wood is the traditional frame material and has higher R-value than steel. It is sometimes clad with metal or vinyl to protect the wood surface and reduce maintenance.

Fiberglass frames, with a core of fiberglass or foam insulation, are making their appearance on the market and offer high insulating qualities as well as low susceptibility to expansion.

Selecting Window Design

For the Millennium Star Home, we are using the awning type and the casement type of window. Casement windows (the type that crank out) and awning windows (they hinge at the top and open at the bottom) are less vulnerable to air infiltration than other styles of windows because they have fewer seams. However, there is no guarantee they will be superior in air tightness.

The horizontal glider type of window and the double-hung type of window have more frame area per opening, making them less energy efficient per square foot of glass area.

MILLENIUM STAR PROJECT

Insulation

Heating and cooling account for about 60 to 80 percent of the average Minnesota household's energy costs. In most cases, increasing the amount of insulation in a home can reduce these costs, but that is only part of it. Improving a home's energy efficiency involves an understanding of the "thermal envelope," the barrier to heat loss and summer heat gain that protects and separates indoor living space from outdoor climate.

Insulation is the heart of the thermal envelope, but it is part of an entire system including siding, sheathing, Sheetrock, and other materials. How well insulation works in resisting heat flow depends on where and how it is placed and on what, and how much material is used.

The type and amount of material used also affects how well the insulation works. Insulation material is rated by two measurements:

- Its R-value, or resistance to heat flow. The higher the R-value, the better the insulation reduces heat flow.
- Its density, or the consistent thickness of the insulating material.

Insulating the Millennium Star Home

Structural Insulated Panels

We used structural insulated panels for some of the wall construction and a portion of the roof. SIP panels are composed of rigid foam insulation sandwiched between two layers of oriented strand board. The oriented strand board provides the structural strength, and no wall studs are used in the panels. The SIPs we used in the walls are 4 by 8 feet and 6 inches thick. The panels lock together, giving the walls an R-value of 23. At the highest point in the roof, we used 10-inch foam panels. It is called a "hot roof," as it is unventilated. SIP construction is energy efficient, strong, quiet, and draft free.

Wall Insulation

To insulate the stud type of wall construction in the other areas of the home, we used three different types of blown-in insulation. Benefits of blown-in insulation include filling the wall cavity 100 percent and the ability to install a higher-density material. Insulation density has a big effect on performance. A high-density insulation will not lose its R-value when temperature drops. For example, a standard 6-inch R-19 fiberglass batt that has low density loses up to 50 percent of its insulating value when the temperature drops below 0°F. Convective currents occur within the wall cavity because of the low density of the material. The extreme temperature difference

MILLENIUM STAR PROJECT

between the inside wall surface and the exterior wall surface is what creates a subtle convective loop within the insulation. With blown-in fiber insulation, the higher the material density, the better the insulating quality.

In the majority of the walls, a dense-pack cellulose was blown in under pressure where a plastic vapor barrier was attached to the wall before insulating. It is made of a recycled newspaper product that has been treated with a fire retardant.

In another wall section, we used a blown-in fiberglass material that also completely fills the cavity. In this case, a netting material is stapled to the wall studs, supporting the material as it is blown into the wall cavity.

Another insulating product used in the Millennium Star home is Icynene spray foam. It is applied by spraying liquid components into the wall or stud cavity. The foam is inert (containing no formaldehyde) and will not biodegrade, shrink, sag, settle, slip, turn to dust, or decline in R-value. Its R-value is 3.66 per inch.

Foam expands to 100 times its size in less than a minute, completely filling the space to be insulated. After it sets, in about an hour, any excess material overflowing the cavity can be easily shaved off with a knife or saw and then reused in areas of an attic for insulation.

The following chart compares density

Insulation Density — 6-inch wall cavity

Regular fiberglass batt	0.56 lbs. per cubic foot
High-density fiberglass batt	1.10 lbs. per cubic foot
Blown fiberglass	1.5 lbs. per cubic foot
High-density blown cellulose	3.0 lbs. per cubic foot

Attic Insulation

In the other attic areas in the house, we used blown-in cellulose 16 inches deep, giving the area an R-value of 50. An important energy detail with construction of an attic and roof is to use a "raised heel" or "energy truss." The raised heel truss allows for full-depth attic insulation all the way out to the perimeter of the attic. Careful attention to detail was paid to providing a solid windbreak at the outer edge of the attic insulation to prevent air intrusion. In ventilated attics, typical soffit and ridge vents were used for ventilation.

Air-Tight Construction

It has only been in about the last ten years that home builders have recognized the importance of building tight and ventilating right.

Houses built in the early to the mid-1900s generally were not insulated and had a high rate of air exchange in the building. Owners of these older houses, particularly those

MILLENIUM STAR PROJECT

in colder climates, experienced problems such as cold drafts, discomfort, dry air, and high heating bills.

Driven by the rising cost of energy and consumer demand for more comfortable and efficient homes, builders began insulating homes, reducing the high rate of air exchange. Improved insulation, along with the introduction of new building materials such as plywood sheathing, thermopane windows, and plastic vapor barriers resulted in homes becoming more airtight, creating a need for mechanical ventilation. The newer, well-insulated homes can have different problems, such as poor indoor air quality, condensation on windows, wet insulation, moisture damage, and wet basements. This is where the term "build tight and ventilate right" comes into play.

The new Minnesota Energy Code now addresses the importance of whole-house ventilation and sealing the interior of the home to prevent structural damage that can eventually lead to mold and mildew.

One of the most cost-effective, energy-saving details in an energy-efficient home is maintaining a continuous seal between the heated and unheated areas. In the winter months, warm, moist inside air is trying to migrate to the outside through the walls and ceilings. Indirect or hidden paths and openings where warm, moist air can leak through the structure are called bypasses. Examples are wire, pipe, and plumbing holes that penetrate exterior walls and attics, spaces around chimneys, soffits and attic hatch doors, gaps around recessed lights, and other areas where warm, moisture-laden air can leak into the home's structure.

Careful attention to sealing these gaps and spaces will eliminate condensation problems that can cause mold, wood rot, and other structural damage to your home. You will also be building into your home durability and longevity.

For as little as $500 to $1,000, depending on the size of the home, you can seal all of the bypasses. The cost includes both labor and materials. Sealants can be various types of acoustical caulks, spray foams, and tape. It makes sense to purchase high-quality sealants that will remain pliable. If the sealant completely dries out, it will break up and expose the original space or gap.

Ventilation

As a home becomes increasingly tighter, it is imperative to provide for indoor ventilation — the exchange of air inside the home. It is separate from attic or roof ventilation and has a very different purpose. It has two essential functions: to exhaust pollutants, moisture, and odors from inside the house to the outside and to bring in outdoor air to mix with the indoor air.

Ventilation is a necessity for controlling the air quality in an energy-efficient home. New energy code rules have recently been implemented as part of Minnesota's building code

MILLENIUM STAR PROJECT

regulations requiring mechanical ventilation in all newly constructed homes.

Siding & Roofing

The use of metal for siding and roofing in home construction continues to increase. Aesthetically, it is clean and sleek looking, and it comes in a variety of shapes and colors. Other advantages of modern steel roofing and siding include easy installation, longevity, exceptional durability, and energy savings.

Roofing

The roofing used for the Millennium Star home is made of 22-gauge standing seam steel. The unique UNA-CLAD patented design is a self-locking system that is fastened without the use of clips and allows for thermal movement. It has an attractive, dark green Hylar paint coating.

A steel roof was chosen for several reasons:

- A steel roof will last up to 100 years, compared to an asphalt shingle roof lasting between 20 and 25 years.
- Steel from a metal roof can be recycled. Asphalt shingles are placed in a landfill.
- Its smooth surface is designed to shed snow and ice, and it is highly resistant to damage from wind or storms.
- Metal reflects most of the sun's rays and cools more quickly at the end of the day. Asphalt shingles absorb energy and retain heat, requiring more energy to maintain a cooler living space in the summer months.

The initial cost of metal is higher. A steel roof usually costs two to three times more than asphalt shingles. Considering metal's increased life span and the fact that it is virtually maintenance free, the life cycle costs of metal are lower than conventional roofing materials.

Siding

The siding on the Millennium Star home is 29-gauge seamless steel in a driftwood gray color. It comes in a continuous rolled stock that is 1½ to 2 feet wide. The lengths of steel are attached to the house with about a 3-inch lap.

Advantages of steel siding include:

- Lifetime warranty
- Low maintenance

MILLENIUM STAR PROJECT

- No rusting or chipping

Heating & Cooling

The Millennium Star Home was designed to use approximately one-half of the heating and cooling energy of a conventional new home. One of the benefits of designing and building a home that is super energy efficient is that not only are the heating and cooling costs drastically reduced, but the cost of installation is lower due to reduced heating and cooling requirements.

To determine the heating and cooling requirements, a heat-loss analysis was completed based on the blueprints, the various components of the structure, the building orientation, and the estimated air leakage rate. The heating requirement for the Millennium Star Home is 24,000 Btu/hour output based on annual degree days of 9901. The projected annual heating cost is $280. Degree days are units used for measuring heating and cooling requirements. Daily degree days are calculated by subtracting the average daily temperature from 65°F. Heating degree days result if temperatures are less than 65°F; cooling degree days if more than 65°F.

For demonstration purposes, we have installed a variety of heating types in the Millennium Star Home. Each type is being monitored so the data collected will show how much each system contributes as a heating source and how the heating energy correlates with the outside climatic conditions.

Five of the six heating types are electric, so we are able to take advantage of Minnesota Power's low cost Dual Fuel interruptible service rate at 3.7 cents per kWh. To qualify for the special low rate, we had to install a nonelectric backup heating source that is capable of continuous operation to meet the total space heating needs of the house. The direct vent gas fireplace with ducted air distribution provides that requirement.

Total installed heat output

	F12	Watts	Btus
Lower level/Basement	102	4500	15,360
Kitchen	153	790	2,700
Dining Room	210	2015	6,880
Family Room & Entry	341	2740	9,340
Master Bedroom	224	900	2,070
Main floor bathroom	112	624	2,130
Left	143	12070	1,710
Totals	2245	12070	41,190

MILLENIUM STAR PROJECT

Energy Efficient Appliances & Lighting

Appliance models on the market today are much more efficient than those of ten or more years ago. And, compact fluorescent lamps are available that use about 75 percent less energy and last about ten times longer than incandescent bulbs. For the Millennium Star home, we installed some of the most state-of-the-art, energy-efficient equipment available.

Here is a summary:

Over-the-Range Microwave Oven

We selected a GE Profile Spacemaker Microwave with sensor cooking controls that automatically eliminate the guesswork on cooking times for a variety of foods, including popcorn, beverages, ground meat, fish fillets, chicken pieces, and vegetables. With 1,000 watts of power, it is one of the most energy-efficient ovens on the market today. Its unique SmartControl System with an interactive display panel allows you to leave audio messages for your family to retrieve and input written messages to remind you of future appointments. A powerful venting system quickly removes smoke and steam from your cooktop below.

Electric Range

Sitting under the microwave is a GE 30-inch free-standing CleanDesign convection range. The energy-saving convection oven cooks 25 percent faster and at lower temperatures than a traditional range. No preheating is necessary. The SmartLogic electronic control delivers more consistent oven temperature for exceptional cooking and baking results. The smooth, seamless surface within the self-cleaning, three-rack oven makes cleanup easier than ever. The patterned glass, ceramic cooktop has a special bridge element that provides for more even heating with cookware.

Side-By-Side Refrigerator

Superior insulation and the automatic energy-saving condenser loop make this 28.2-cubic-foot capacity GE refrigerator quiet and economical. The light touch dispenser on the door front delivers crushed ice, cubes, and Culligan filtered, chilled water. Adjustable shelves, slide-out freezer baskets, and deep doors that can hold gallon cartons of milk provide for easy access and convenient storage.

Laundry

Revolutionary to the task of washing and drying clothes is the front-loading ASKO Laundry Care System. We installed the Swedish-made ASKO washer and dryer in the Millennium Star to demonstrate their remarkable ability to do a better job of cleaning and drying with a fraction of the water, electricity, and detergent as in ordinary machines.

MILLENIUM STAR PROJECT

Washer

The washer uses as little as 5.7 gallons of water per load compared to 44 gallons for most conventional top-loaders and 28 gallons for domestic front-loaders. They are smaller in size than conventional machines, yet they still handle big loads — up to 15 large towels. One of the most unique features of the washer is that it connects only to the cold water line and uses an internal heater that can reach temperatures as high as 205°F.

Dryer

The ASKO dryer spins clothes at speeds as high as 1,600 rpm — about twice as fast as top-loaders and most front-loaders — so they can extract at least 40 percent more water than conventional machines for a much shorter drying time. This microprocessor-controlled dryer has precise heat and humidity sensors to prevent overdrying and reduce wrinkles.

ASKO Dishwashing System

Like the ASKO washer and dryer, the dishwasher is hailed as a world leader in innovation and quality. It uses less water, energy, and detergent than other models, and its larger capacity saves you time in the kitchen. You will not find a quieter model on the market today.

Courtesy Dean Talbott, Minnesota Power, serving 141,000 electric customers in Northeastern Minnesota and Northern Wisconsin.

MILLENNIUM STAR
MINNESOTA POWER'S MODEL ENERGY-SAVING HOME

Gas Fill in Windows

A modern window has two panes of glass about an eighth of an inch apart. Under normal circumstances, this space fills with air. A gas-filled window replaces the air with argon or krypton. This works because air is composed of 78 percent nitrogen and 21 percent oxygen. This works because air is composed of 78 percent nitrogen and 21 percent oxygen. Since argon (atomic weight 39) is more dense than nitrogen (at 14) and oxygen (at 8), it prevents transference, or thermal conductivity. Check with the manufacturer, seller or building contractor to determine if your windows are argon-filled, and how much argon was used. More in this case is better.A high-quality, gas-filled window will lose less than 2 percent of argon per year. This means it will take 50 years for your gas-filled window to lose its thermal properties, and modern windows have a "life expectancy" of about 20 years.

Window Coatings

Window coatings are either soft or hard. Soft coatings, tin or silver, are applied to the insides of the panes, where the gas fill is located, sealed between two protective oxide layers. A hard coating is made of a tin-oxide alloy and applied to the outside surface of the glass. Hard coatings tend to be more durable than soft coatings and also are less expensive in terms of the whole window. Coatings also can be applied to polyester films and sandwiched between the glass panes to provide a third air barrier. Tin's atomic weight is 118. The atomic weight of silver is 107. If you read the previous paragraph on atomic weights, you will understand how low-e coatings work to prevent air conduction.

Spacers

Spacers between the dual panes of glass in a modern window were originally made of metal. Metal is highly conductive, which is why we use metal in pots and pans for cooking food. These metal spacers are being replaced by non-conductive substances, such as rigid foam, fiberglass, or wood.

Frames

Wood is best, but aesthetic issues with wood such as rotting, peeling, and paint fading have forced manufacturers to coat wood frames with vinyl. Well-designed vinyl and fiberglass frames can be even better than wood, particularly if hollows in the frame are filled with insulation. These, however, are more expensive than vinyl-clad wood, and cheaper models are not as thermally effective. Aluminum frames, even those with thermal breaks — spaces where the metal is separated by polyurethane — may be less efficient, and the stability of the frame is further compromised by the "breaking" process.

A window's frame can comprise up to 25 percent of the window's total surface area. Choosing a window with a smaller frame is one way to ensure energy efficiency because you will be using more natural light rather than artificial, electrical-using interior lights. However, if the window you can afford has an aluminum frame, smaller frames will result in less durable windows. Fiberglass frames offer both insulation and strength in lower-profile frames and are a good economic choice between vinyl-clad wood and aluminum.

Wood windows are easy to handle, absorb sound well, and tend to be easy to repair, though they do require periodic painting. Major advances have been made in all-wood window manufacture, bringing their energy efficiency and ease of installation in line with the best-quality vinyl windows. The only problem is the wood itself; timber harvesting remains a sensitive, ecological issue.

If you buy wood windows, be sure you are buying Forest Stewardship Council-certified wood. The Forest Stewardship Council (FSC) is a nonprofit foundation that certifies sustainably harvested wood. The cost of such wood is about 10 percent higher than other woods, but the Forest Stewardship Council is the only certifier recognized by the Green Building Council and other environmental organizations.

For more information on Forest Stewardship Council-certified wood, go to **www.fscus.org** or the Green Building Council (**www.usgbc.org**), or go to Greenpeace and read information on selecting certified wood (**www.greenpeace.org/usa/news/how-to-find-and-avoid-toxic-vi/certified-wood-programs**).

New Window Alternatives

If you cannot afford new windows but still want to reduce your energy costs and your carbon footprint, you can do a number of things to improve your existing windows' energy efficiency. Most of these options require you to retrofit your windows with an additional covering of some sort. If you are a fresh-air enthusiast and do not live in an area where you can leave your doors open to provide outside air through the screen door, you may want to try these ideas on a few windows at first. Once you see their advantages in energy savings, you will be more comfortable proceeding. I have grouped them by order of cost, from the most expensive options to the least expensive alternatives.

Storm Windows

Buying storm windows for your home is vital if you live in an older home with single-pane windows in an area where winters are cold. In fact, if you have an older home in the Midwest, it probably came equipped with storm windows. If it did not, you may be able to find storm windows at a salvage yard, on **CraigsList.org**, or by advertising in your local newspaper. People who are upgrading to new windows often have storm windows they want to get rid of.

Storm windows were, traditionally, installed on the outside of the house. Nowadays you can buy interior storm windows, and these are useful if you want to increase your energy efficiency but do not want to obstruct the windows in your classic, Victorian home. Interior storm windows can be left in place year-round or can be removed in summer for better ventilation. They also offer the same low-e glazing available in new, energy-efficient windows and the same ultraviolet-resistant film. Acrylic versions, however, do not withstand scratching, and the acrylic may discolor after a decade of use.

Storm windows act like double-pane windows — the air trapped between the inside window and the storm window absorbs heat, or cold, from the air inside the room and slows the transference of air into a room. However, if the storm window does not fit the window opening tightly, or the caulk around the edge of the panes is old and cracked, a storm window will be of little benefit in keeping cold air out. You can use weather stripping to improve the fit of the window, and you can scrape the old caulk and apply new. Both jobs are time consuming but quite inexpensive; a roll of 3M "V" strip is about $10 and will seal three medium-sized windows. Caulk is $5 per tube and will renovate seven- to ten-year-old windows. If you cannot afford to buy storm windows for the whole house, start on the north side and work your way around, from east to west.

You also can buy a storm-window kit for about $15 per window, but these are nothing more than vinyl and double-edged tape, which you can purchase and install yourself for much less.

Window Insulating Kits and Window Films

There is a difference — insulating kits allow you to apply plastic to the inside or outside window frame. 3M makes an excellent window- insulating kit that comes packaged with double-faced tape. For about $18, you can cover three good-sized windows with this crystal-clear film and keep your room toasty all winter.

You can buy other types of window-insulating kits, or film, at home improvement stores, such as Lowe's or Home Depot. Sheet plastic is measured in millimeters, or mils. A roll of 8-millimeter, crystal clear, ultra-violet-stabilized, reusable plastic, will cost $75 to $100 for a 72-inch wide roll about 75 feet long. You also can find window films online. Some are clear, some frosted, and the thickness of the plastic can vary from two to 7 mils. All window films are polyester. Insulating window films are commonly applied to a frame with double-faced tape. Glare-reducing, static-cling films are applied to the insides of windows, block ultraviolet rays, and can be removed and reused. Privacy films can be frosted, or one-way; that is, you can see out but no one can see in. We are talking exclusively about insulating films, which are not even strictly films, because they do not attach to the glass.

To install window insulation film, first measure the windows you plan to cover and allow 15 percent for waste. Buy rolls that most closely correspond to your window widths, or buy film as wide as your widest window, if it is twice as wide as other windows. Then clean your windows thoroughly, including the frames. Apply the tape around the frame, then apply the film to the top and then bottom strips of tape, working the film toward the corners. Seal the sides last, being careful not to overly handle the plastic, as this may leave fingerprints.

When you are finished, you can take out the wrinkles with a hair dryer. This film can be removed and stored for use the following winter. The tape is not removable, but if you save the strips that covered it, you can put these back in place to keep your double-faced tape "sticky" for another year or so.

Insulating films, as opposed to kits, are applied to the glass and can cut energy use significantly because of their low-e coatings. Insulating films commonly have a water-activated adhesive on one side. More-expensive films provide additional benefits, such as glare reduction and privacy. Good-quality films have a scratch-resistant surface for long wear. Films are more difficult to apply than insulating kits but are well worth the investment of your time. For more information on window films, go to **www.lowes.com/lowes/lkn?action=howTo&p=/Energy/WndFlm.html.**

Once you get your windows airtight, you will have other problems besides high energy bills and an equally large carbon footprint. Modern homes, which allow less air transference than their older relatives, are prone to mold, mildew, condensation on windows, and "sick-building" syndrome; indoor air pollutants do not have a chance to escape the way they once did, and these pollutants are more prevalent now than ever before. They exist in our carpets, the paint on our walls, the fabrics in our furniture, the chemicals we use to clean, and our cosmetics and tend to build up to toxic proportions. According to the Environmental Protection Agency, the primary cause of sick-building syndrome is lack of adequate ventilation.

Sick-Building Syndrome: Remedies for Airtight Homes

You may not encounter the problem in an older home, even if you seal all the windows. Opening and closing doors lets in fresh air, as do your hood fan and bathroom fans. If you do not have hood or bathroom fans, you may notice symptoms of illness as condensation on windows first. When you do, you will need to think about ventilation, particularly if you have a fireplace or wood stove, which can suck a great deal of usable air from your home, as can other open-combustion sources, such as gas dryers and gas hot-water heaters.

Whole-house Fans

A whole-house fan is a simple and inexpensive method of exchanging bad air for good and can supplement your home's cooling system in summer. Using a whole-house fan in summer can cut cooling costs by $6 per day. Installed in the attic, it may move about 4,000 cubic feet of air per minute. These fans cost about $300 for an Energy Star-rated fan and are relatively simple to install. Unfortunately, they do not work as well in the winter when you are heating or during periods of heavy air-conditioner use, because there is no way to insulate the unit and prevent it sucking in, and then blowing out, heated or cooled air.

Gas forced-air furnaces bring in cold air from the outside to support combustion, and you can run your furnace on fan-only setting during the summer to provide adequate ventilation. In the winter, all the air coming in is used to support combustion and is not available to your home's air supply. You can, however, install an air cleaner on your furnace and run the furnace on fan only in the summer. In the winter, the air cleaner will work whenever the furnace is blowing heat.

Air Cleaners

If you have hot-water heat, you may need to invest in at least one good air cleaner. There are three basic types of air cleaners: mechanical, electronic, and ozone generating.

Mechanical filters force air through a screen, or filter, that traps particles, such as pet dander, and particulates, such as cigarette smoke. The best-known mechanical filter is the high-efficiency particulate air (HEPA) filter, which is a type of filter, not a brand name. To qualify as a true HEPA filter, a device must be able to capture at least 90 percent of airborne particles larger than 0.3 microns in diameter. Check the filter specifications before buying; just because it says HEPA on the box does not mean it works accordingly.

Mechanical filters, or HEPA filtration devices, now come equipped with ultraviolet light fixtures. The Centers for Disease Control recommends the use of ultraviolet light with HEPA filters to increase indoor air quality in homeless shelters. Ultraviolet light controls microbial growth. It is the ultraviolet component of sunlight that makes colds and flu less likely in the summer. Bacteria and viruses die within a few minutes at most when exposed to ultraviolet light. In fact, ultraviolet light may be a final line of defense against mutating, antibiotic resistant strains of tuberculosis and Legionnaires' disease, a type of pneumonia associated with warm-water environments and crowds and first identified at a 1976 American Legion convention in Philadelphia. A study by Air & Waste Management Associates found that use of ultraviolet light in association with HEPA filtration reduced bacteria by 80 percent in a 10-by-10-by-30-foot closed environment. These combined UV-HEPA filters come as stand-alone units or can be installed in forced-air furnaces.

Electronic filters, also known as electrostatic precipitators or ESP units, use electrical charges to attract and deposit allergens and irritants. Some of these filters have actual collecting plates inside that can be wiped down to remove the trapped products of air pollution. Others do not, and the particles stick to surfaces in your home, necessitating plenty of dusting. These air cleaners generate negative ions, and, while cleaning air better than mechanical filters, have been linked to dangerous levels of ozone, according to *Consumer Reports*' April 2005 issue (**www.aircleaners.com/sharperimage2.phtml**). Of brands tested, the Sharper Image Professional Series Ionic Breeze Quadra SI737 SNX, Brookstone Pure-Ion, Ionic Pro CL-369, IonizAir P4620, and the Surround Air XJ-2000 were all ineffective as air purifiers and produced enough ozone to trigger asthma attacks or intensify allergy symptoms in sufferers. Ozone also can deaden a sense of smell and may cause permanent lung damage.

Cleaning the Air with Plants

Green plants do a remarkable job purifying indoor air. They take in carbon dioxide and give off oxygen. Plant species vary in their uptake of different chemicals but were found as a whole to remove volatile organic compounds from carpet, upholstery, and paints and alleviate odors from mold, mildew, and cooking. Odors are particulate.

Plants vary in their ability to remove pollutants. Spider plants, philodendrons, and pothos were most efficient at removing formaldehyde; gerberas and chrysanthemums were most effective at eliminating benzene, a known cancer-causing agent. Peace lilies also remove carbon monoxide and are one of the easiest plants to grow, requiring adequate moisture but relatively indifferent to lighting. A north window in winter provides sufficient light to keep a peace lily happy.

For an average home of 1,800 square feet, one would need 18 large plants, or one plant per 100 square feet. Large is defined as more than a foot in diameter. Below is a list of air-purifying plants:

- Chinese evergreen (Aglaonema)

- Burn plant (Aloe barbabensis)
- Spider plant (Chlorophytum comosum)
- Chrysanthemum, or mum
- Dumb cane (dieffenbachia)
- Pothos (epipremnum)
- Ficus
- Gerbera daisies
- English ivy (Hedera)
- Philodendron
- Peace lily, or mauna loa (spathiphyllum)

Aloe, mums, dumbcane, ivy, and philodendrons are toxic if eaten, so keep them away from pets and children. Dumb cane, if ingested, can cause mouth irritation, burning, drooling, vomiting, and difficulty in swallowing. If you suspect your child or pet has chewed or ingested this plant, call your doctor, veterinarian, or poison-control center immediately.

Cleaning with Light

There are a number of light bulbs on the market that claim to purify air. One, the anionic light bulb, is a compact fluorescent with a built-in ionizer. It produces significantly less ozone than an air cleaner and also provides illumination. It saves up to 80 percent of energy costs; lasts for up to 10,000 hours, or eight times longer than an incandescent bulb; provides 100 watts of illumination; and can remove smoke, pollution, pollen, dust, dust mites, odors, and pet dander from surrounding air. It promises relief for allergy sufferers and insomniacs, or those who have frequent headaches or fatigue. The anion light bulb can aid in the body's absorption of oxygen and increase serotonin levels to reduce stress and anxiety, makers claim. There are a number of manufacturers of these bulbs, and the average cost runs from $17 for one to $30 for two.

Ultraviolet light also cleans air; however, these light bulbs cannot be used in an occupied room. Exposure to ultraviolet light will cause eye irritation and reddening of the skin. You might want to install them in a kitchen and use them as a night light, but you will have to educate other family members on their dangers.

The newest invention in this area is a titanium-coated bulb, available as a compact fluorescent, which claims to eliminate odors. One reviewer who tried some agreed that they live up to their advertising. Apparently, when exposed to light, the titanium dioxide coating creates oxidizers that break down and dissolve odors. Another version of this light uses titanium dioxide and ultraviolet light to trap and kill bugs. Called "fresh" bulbs and sold under the brand name Fresh2, these titanium dioxide-coated bulbs do not present any known health hazards. Titanium dioxide is not a carcinogen, mutagen (causing mutations), tetratogen (interfering with the development of a fetus), or a trigger for contact dermatitis and often is used as a base pigment in foundation, or skin makeup, as well as a white base element for colored paints.

Flooring

Floors are an integral part of a home. They also can be a huge source of indoor air pollution. Most floors are covered with carpeting, vinyl, or wood. Even the subfloor is a combination of wood fibers, or chips, and petroleum-based glues, if it is either plywood or particle board, as most subfloors are.

Modern carpeting is made primarily of synthetic fibers, coated and treated with stain and water-resistant chemicals. Ninety-seven percent of the carpeting we buy is made from nylon, olefin, polyester, or acrylics. These man-made compounds use petroleum and about 100 other chemicals in their manufacture, many of which are volatile organic compounds. These volatile organic compounds provide stain and water resistance; repel fungi, bacteria, and insects; stabilize colors and fibers; and keep the tufts attached to the backing. All are toxic.

Nylons are condensation copolymers formed by reacting equal parts of diamine and dicarboxylic acid, which is a component cause of acid rain. Polyesters are made from polyethylene terephthalate; we have already covered the dangers of phthalates. Other forms of polyester include synthetic polyesters, such as polycarbonate and polybutyrate, which is short for polybutyrate adipate terephthalate. Acrylics are made from N-dimethylformamide or aqueous sodium thiocyanate. Degrading dimethylformamide often has a fishy smell, particularly when it gets hot. Thiocyanate salts are made from a reaction between cyanide with elemental sulfur.

Carpets can emit these toxic chemicals for up to four years after installation. In damp weather, according to one study, carpets form and exude alcohol esters. In sunlight, outgassing is stepped up, as heat increases the reactivity of these volatile organic compounds, both to each other and to chemicals in the environment. Many people are allergic to new carpeting, and the symptoms can include headaches, dizziness, slurred speech, brain fog, memory loss, chills and fever, anxiety, numbness, depression, and seizures. To some people, this carpet allergy feels like the flu, causing sore throats, joint pain, and coughing. Many people have acquired multiple chemical sensitivity illness (MCS), which can be disabling and last a lifetime.

After it has worn out, we throw the carpeting in a landfill. Three percent of landfill capacity is occupied by old carpet, which takes about 100 years to break down. When it does, all the various chemicals that have not already entered our soil, air, and water finally do so, completing their mission of pollution, whose end result is to make us ill and the planet uninhabitable.

Vinyl flooring is commonly made of polyvinyl chloride. Polyvinyl chloride has been associated with an increase in asthma, particularly among children. Vinyl flooring is made with chemical softeners (dangerous phthalates) to keep it pliable.

Laminate wood floorings are made of bonded layers of wood fibers constituting the image layer, and glues and surface sealers making up the wear layer, with all layers bonded using either direct pressure or extremely high pressure and temperature. This is called "engineered wood" and can come as planks or tiles, the former more natural-looking, the latter making it easier to replace damaged areas. They are priced by the square foot, from as little as $2 to as much as $20, and most contain some level of formaldehyde as a chemical component of the adhesive. Other volatile organic compounds might also be present, such as petroleum-based resins. These laminates, when worn, are either burned or sent to a landfill, causing further problems with pollution.

If you live in an older home, your subfloor and floor may both be solid wood. In that case, you should strip your floors of all covering, sand them, and refinish them using a vegetable-based wood sealer. Please rent an industrial air filter when sanding, or the old residue of varnish and stain on your floor might compound the problem you are trying to solve. If you are not lucky enough to have real wood floors but you do have modern carpeting or vinyl that is outgassing and causing indoor air pollution, you can choose from one of the options below.

Natural Carpeting

The most common natural carpet fiber is wool. Our grandparents had it and replaced it with more modern fibers. We are rediscovering its benefits. Wool is warm in the winter yet surprisingly cool in summer. It is more resistant to matting and wear than synthetic fibers; cleans easily because each fiber is covered by a thin membrane, making it naturally stain resistant; it traps dirt in the dense, uppermost layers so soil is easily vacuumed away; is flame-retardant due its high moisture content and does not give off toxic fumes even when burning; is nonallergenic and does not support bacterial growth; and purifies indoor air of pollutants, such as formaldehyde, nitrogen dioxide, and sulfur dioxide, by locking them deep in each fiber's core. Wool can absorb up to 30 percent of its own weight in water, helping regulate the humidity in your home, eliminating those nasty shocks you get walking across synthetic carpet and then touching someone or something. Wool also is a natural sound insulator, making rooms quieter.

Cotton, another natural fiber, is found in linens and blankets but makes up only 1 percent of the carpeting marketplace. This may soon change. Carpet manufacturers, responding to the growing "green" movement, are beginning to realize that cotton could provide a significant portion of America's carpeting demand. Twenty percent of fiber production goes into carpet manufacture; this translates into 300 million pounds of cotton. If cotton carpeting attained only a 5 percent market share, this would be 15 million pounds of cotton.

Cotton rugs are available, and a strong demand for natural fibers has pushed this market segment to 28 percent, up 22 percent from five years ago. Small, washable cotton area rugs come in woven, braided, and even flat styles. Because cotton is not crush resistant, it is best blended with wool to attain this attribute. Linen, another natural fiber made from flax, is seldom seen in carpet except as a blend of ingredients. Linen is soft, durable and flexible, but the resource is increasingly scarce.

Other fibers that can stand alone or be woven with wool include jute; sisal, made from the agave plant; and coir, made from fibers from the husk of a coconut. These fibers are durable, chemical free, and naturally flame-and-insect resistant, and sisal has remarkable sound-absorbing properties.

Unfortunately, most are coarser than the carpets we are used to and thus have not gained popularity.

For more information, go to **www.greenhome.com/products/home_furnishing/rugs_and_carpets**.

Natural Soft Flooring

Natural linoleum is made from linseed oil pressed from the flax plant, then mixed with pine resin, wood sawdust, cork powder, limestone dust, and jute. Natural pigments, such as ochre from clay or rusty red from iron oxide, are used to color it. It can be laid almost anywhere a soft, resilient floor is required and can be used for countertops and desk surfaces. It comes in tiles and sheets.

The development of plastics rendered it almost obsolete for 60 years, but natural linoleum is making a comeback. Armstrong recently started marketing its own brand of linoleum, imitating European manufacturers who never stopped making the product.

Natural linoleum, as opposed to petroleum or asphalt products wearing the same name, has many advantages over vinyl. Natural linoleum becomes stronger over time and lasts 40 years compared to vinyl's 20. Linseed oil has natural, antibacterial properties, as does limestone, and these properties are retained in the finished product. Natural linoleum is easy to maintain, can be dry mopped or wet mopped with mild soap and water, and requires only an occasional application of sealant to keep it looking new. Unlike vinyl, linoleum's color runs entirely through its thickness, so scratches and dings can be sanded and sealed to restore that "new" look. Natural linoleum is "anti-static" and repels dust and dirt. Linoleum does not outgas volatile organic compounds, though linseed oil fumes will be apparent both during and after installation and will continue until the linoleum's surface "cures" in about a week. Being a natural product, it is not rot-proof and should be properly caulked and sealed when used in damp areas, such as bathrooms. When it reaches the end of its usable lifetime, it can be recycled, either by being ground and made into new linoleum or ground and spread in landfills, where its natural ingredients degrade safely without toxic residue. Manufacturing costs are small, and the process is very resource efficient; scraps are fed back into the process, generating zero waste. Most real linoleum is still made in Europe, so shipping costs account for most of its price.

Forbo, currently the best-known linoleum producer, offers products in a wide range of colors and styles (**www.forbo-esd.com/framework/DesktopDefault.aspx?menu_index=0&menu_id=1**). Naturlich also offers natural linoleum (**www.naturalfloors.net/linoleum.html**). Armstrong has begun marketing its own line of natural linoleum, in response to consumer demand for the product (**www.armstrong.com/resflram/na/linoleum/en/us/**).

Costs are comparable to high-quality vinyl flooring, about $30 per square yard, though installation costs are somewhat higher. Most experienced flooring contractors agree that natural linoleum is not as flexible as vinyl, requires more attention to detail, and requires more caulking and sealing of seams. It is an exacting job, and some flooring experts may refuse to install it due to lack of experience. Linoleum has to be "air-cured" for a week before installation to make it workable. Even then, it is not appropriate for damp basements.

Hardwood Flooring

Nothing compares to a real wood floor. Lumber, once a living thing, retains its history in every knot, burl, and grain, revealing a unique character that man can imitate but never duplicate. Wood is durable, forgiving, and ages gracefully but will permit any number of refinishing jobs and options without losing its charm.

We take it for granted, but that is beginning to change. Environmentally responsible individuals and groups are beginning to question the endless, and sometimes reckless, harvesting of this remarkable resource. People are becoming aware that sustainable forestry, a boon to trees, also is a benefit to humanity. Our responsible choices will mean that wood is available to future generations without depleting a resource that assumes half the responsibility for providing us with the oxygen we need to breathe. For more information on the Forest Stewardship Council, certified woods, and green building, see the section on new windows, or go to **www.fscus.org**. Do this before you buy; other wood certification systems are either government or industry sponsored and do not guarantee sustainably harvested wood.

Real wood floors are expensive but cost effective in terms of their lifetime. Carpeting costs roughly $3 per square foot, not including installation or pad, will have to be shampooed every few years by renting a shampooer or hiring a carpet-cleaning firm, and lasts about 15 years in normal, residential wear. Wood costs $4 per square foot uninstalled, might need sealing every other year and sanding at half its lifetime, can be mopped rather than shampooed, and lasts 30 to 50 years. In fact, you might tire of the wood before it has to be replaced.

Resist the temptation to cover your wood floor with carpet. Wood does not contribute to allergies, support mold or mildew, harbor dust mites, emit volatile organic compounds, or retain up to 30 percent of its original weight in dirt and spills by the end of its usable lifetime. The only real problem you may encounter with a hardwood floor is a tendency for the planks to shrink or warp in the presence of moisture, which causes squeaking and crackling. You can avoid these problems by controlling the humidity in your home and mopping carefully with an almost-dry mop and a solution of Ivory soap, which will not damage hardwood-floor finishes.

Demand real wood; do not accept substitutes. An "engineered" wood floor is a composite of wood layers bonded using petroleum-based glues and coated with ultraviolet-resistant sealers, all of which emit volatile organic compounds. Laminates present the same problems, intensified by the extensive laminating process. If you cannot afford a new wood floor, consider "reclaimed" wood, and read the section on salvaged floors.

Other Natural Flooring Choices

Cork flooring, made from the bark of a tree, has a unique cellular structure, providing a flexible yet durable walking surface that is naturally anti-microbial, deadens noise, and does not transfer cold or heat from adjacent surfaces. Cork is relatively cool in summer and amazingly warm in winter. Cork can be pale bisque, soft rose, or muted walnut and provides a low-impact surface that comforts aging backs, feet, and joints.

Cork is an excellent environmental choice because harvesting cork does not kill the tree. Cork trees live an average of 200 years, and the bark, with its thousands of air-filled chambers, is impermeable to water and gases, contains natural antiseptics, and can even be eaten. Manufactured into flooring, in a process little more chemical-intensive than minor gluing and pressure-treating, cork is then sealed with ultraviolet-cured acrylic. At the end of its lifetime as your floor, or in a wine bottle or golf ball, cork can be recycled into more flooring or golf balls.

Neutral in color, with distinctive yet nonintrusive patterns, cork blends with any décor, from casual to formal, making its statement without shouting. It will not dent and mar like wood flooring, and any cuts or grooves that do occur are minimally apparent because of its pattern. It is very lightweight; is not slippery like wood flooring; provides good traction for area rugs, including cotton rugs; and is impervious to food and oil spills. It is, in my opinion, the most environmentally sound flooring choice a family can make. Natural cork, as opposed to highly recycled cork — which contains more volatile organic compounds — costs between $4 and $6 per square foot. This is more than hardwood but is well worth the investment.

Bamboo is another excellent environmental choice for floors. Bamboo grows quickly, and, once harvested, regenerates from its roots, so the supply is sustainable. In fact, bamboo has to be harvested; if it is not, it stops growing and eventually collapses from its own weight. Established bamboo plants are almost impossible to kill, and harvesting is government regulated. The manufacturing process is simple: cutting, boiling, milling, some gluing, and a final sealing with urethane. Some planks are carbonized, or pressure steamed, to make them darker. There are some volatile organic compounds involved, but these are six times less than European standards allow; European standards are stricter than American standards. Bamboo is harder than red oak or maple, withstands wear, and may be warranted to last 20 years.

Bamboo has to be "air-cured" for three days before it can be installed, in an ambient humidity of about 50 percent. It can be glued with 100 percent urethane glue, stapled, or nailed. If used on concrete floors where there is a chance of dampness, lay an acrylic or plastic sealer before installing. For more detailed installation information, go to **www.greenbuildingsupply.com/utility/showArticle/?objectID=54**.

Maintenance is fairly simple. Vacuum often enough to keep sand and grit from damaging the finish, use furniture-leg protectors to prevent punching holes in the bamboo's surface, and use a dehumidifier or humidifier to maintain a constant 50 percent humidity in your home. Mop bamboo with an almost-dry mop and a solution of Ivory soap as needed, buff away scuffs with a Magic Eraser, seal your bamboo with a urethane finish when the original finish shows signs of wear, and wipe up spills immediately.

Bamboo is naturally antibacterial, edible, renewable, and biodegradable. It can safely go into landfills or be reprocessed into fibers for clothing, furniture, or household products, such as table mats and bowls. In my opinion, it is second to cork as an environmentally sound choice. It also is somewhat cheaper than cork, at about $3.50 per square foot for premium-strand, natural bamboo. Although not as colorful or as intricately patterned as cork, bamboo fills the need for a wood-like floor without the disadvantages of hardwood.

Demand for bamboo flooring has resulted in some unscrupulous manufacturers using bad harvesting practices, excessive chemicals, and improper procedures in producing their flooring. You can prevent them making a "quick buck" by choosing an established manufacturer and asking for Material Safety Data Sheets on your product. The Forest Stewardship Council has not yet stepped up to help regulate bamboo, but in 2005 the Chinese government agreed to support sustainable forestry. To view this report, go to **www.wwfchina.org/english/downloads/newsletter/wwfnewsaprilmay05.pdf**.

Tile Floors

The difficulty with so many products that are advertised as eco-friendly is that while they do recycle products previously destined for landfills, the recycling process itself — with its intensive manufacturing cycle and use of glues and sealers — adds to the carbon footprint of the consumer and increases the load of volatile organic compounds in the typical home. Recycled carpet is a prime example: The "bad" chemicals that were in the old carpet persist, and interact, with additional, new, and equally bad chemicals in the remanufacturing process. The result is a toxic load in both the manufacturing cycle and the consumer cycle, leaving the consumer to weigh the benefits of sustainability against the negative effect on your family and the environment. My advice is to choose environmentally friendly where you can afford it and sustainability where the benefits seem to outweigh the drawbacks.

Glass escapes these parameters. Glass is made from sand, or "natural" glass, such as volcanic magma. Recycled glass contains no extraneous ingredients to make it glass again. Heat, rather than chemicals, is the primary catalyst. Melting recycled glass is good for glass-melting furnaces. Melting recycled glass is good for glass-melting furnaces because recycled glass requires less heat and therefore less energy, thus prolonging the life of the furnace. Substituting cullet for raw materials also can reduce emissions into the atmosphere, according to an article called "Recycling, Glass," written by C. Philip Ross and published in the *Kirk-Othmer Encyclopedia of Chemical Technology* by John Wiley & Sons, Inc.

In 1980, glass made up 10 percent of the composition of landfills. By 2003, that figure was down to 5 percent, or about 2 million tons of glass that got recycled into everything from new bottles and jars to wall and floor tiles. Recycling glass uses 40 percent less energy than making new glass. Recycling glass into tiles is environmentally effective and technically advantageous. According to a recent report, the strength of recycled glass tiles, made according to sound engineering specifications, is as good as, and often better than, American Society for Testing and Materials (ASTM) requirements for similar materials.

Before the environmental movement, most consumers viewed products made from recycled materials as poor, though not always cheap, substitutes for their originals. This is not true anymore. The environmental movement of the 1960s has merged with the "second-hand chic" mind-set of the 1980s to produce a generation of consumers who see and expect real value and solid construction in recycled products. Glass tiles live up to their expectations.

There are several recycled glass tile manufacturers on board with these expectations. Recycled Sandhill Industries of Boise, Idaho, buys only broken/discarded glass from local window manufacturers to create recycled tiles. Eco-Friendly Flooring of Madison, Wisconsin, and Bedrock Industry of Seattle, Washington manufactures 100 percent recycled-glass tiles. Interstyle Ceramic & Glass Ltd. of Canada carries three lines that are 100 percent postindustrial recycled.

Not all recycled tiles are strictly glass based. Wausau Tile of Wausau, Wisconsin, makes a terrazzo tile that substitutes glass chips for marble or stone, as does Icestone of New York. VitraStone of Colorado uses about 77 percent recycled glass, adding ash and a proprietary blend of natural-mineral ceramic cement. Terra Green Technologies uses about 60 percent recycled glass and a ceramic stabilizer. All make remarkably beautiful tiles that will spotlight your floors and your eco-sense.

There are other tiles besides glass. Ceramic tile may be either porcelain or clay and is sometimes called paver tile. Ceramic is extruded, while porcelain is formed by dry-pressing. Quarry tile, or quarry pavers, are made one at a time and are thus more irregular, may be unglazed, and are made from shale or clay via an extrusion process. The raw materials for these tiles consist of refined clay, shale, natural minerals such as feldspar, and chemical additives needed for the shaping process. The ingredients are mixed, sometimes with water, and the result is either wet-milled or ground up in a ball mill. The resulting powder is then pressed into the desired tile body shape or, if wet, extruded into the desired shape. The glaze, or sealer, is either sprayed on when the tile is complete or sprinkled on the unbaked tile. The baking process forms the glaze. Tiles may need to be baked once or several times. Tiles are labor intensive to produce, but the only pollutants generated are lead and fluorine compounds, most of which are controlled by advances in technology. A greater problem is the recycling of wastewater and sludge, because particles of clay that do not make it into tile remain suspended in wastewater. These issues are also being resolved by returning waste by-products to the production stream.

There are hundreds of names for tile, including terrazzo, mosaic, and quarry, but all true tile is made from the same basic ingredients and formed either mechanically or by hand in molds. All tile is baked in ovens, or kilns, just like pottery. Floor pavers that are not baked are stone or aggregates of stone and minerals. Some are called cement-body tiles; that is, tiles made from a mixture of sand and Portland cement whose surface is finished with chips of marble or other materials and sealed.

Stone Flooring

First, the elemental differences between types of stone: Granite is formed deep in the earth, of minerals that crystallize at extremely high temperatures and pressures, and is virtually impervious. Granite is an igneous rock. Marble and its relatives, such as slate, schist, and travertine, are formed at the bottom of bodies of water and begin their lives as sediment, which is composed of the skeletons and shells of marine animals, plus a little silt and vegetable matter. In a million years, this sediment becomes stone. Because their main component is calcium, marble and its relatives are affected by acetic acid, as found in vinegar and orange juice. Marble is a metamorphic rock. Limestone is a sedimentary rock.

Granite is the second-hardest natural substance known to man; the first is diamond. Granite requires a huge expenditure of energy to get out of the ground. Blasting was a common practice for decades. More recently, techniques such as drilling and wire-sawing have been used to reduce the waste that is an inevitable result of explosives.

My objection to stone is that it is not a renewable resource. Granite, limestone, and marble were all millions of years in the making. They represent a significant investment and will add to a home's value, but the weight of granite and marble sometimes require reinforcing the average residential floor. In spite of those objections, granite may be the most durable, heat-resistant, chip-resistant, and bacteria-resistant substance on the planet. At about $5 per square foot, not including supplies or reinforcement for your floor, granite is on the upper end of affordable. Maintenance costs are almost nil; a sealer once every four or five years is all granite will ever need.

Marble weighs slightly less than granite and costs about the same. Slate and travertine are formed in layers and quarried the same way. They weigh slightly less than marble and are less durable. They cost less than marble, at about $2-3 per square foot, but are somewhat prone to shearing and require considerable care installing. Used to build the Colosseum in Rome and commonly used in modern architecture as a cladding, travertine has recently become popular as a flooring alternative. Travertine, with its classic, creamy white color and "soft-stone" appearance — in which many natural pits may be filled with grout and sanded — transforms an ordinary kitchen floor into an Italian piazza, filled with light and warmth.

However, my original objection stands firm; stone is environmentally friendly but not a renewable resource. Unless there is enough to go around, I vote against it.

Salvaged Floors

If you cannot afford a new wood floor, consider "reclaimed" wood. By this, I mean salvaged floors, not wood recycled by mechanical and chemical processes into "new" wood.

With the advent of environmental awareness, salvage yards are springing up everywhere, offering homeowners the option to reuse someone else's old floor, old cupboards, old bricks, and even old ceramic tile. These floors can come from such diverse sources as homes, offices, abandoned factories, textile mills, barns, and breweries and can represent old flooring, or the joists, timbers, and supports used in construction. This wood, commonly called "antique" or "old-growth" wood, is more than 100 years old and may be stronger than the new-growth wood being cut from the forests today. With its dark patina, intricate grain, and occasional nail holes or other scars, this wood lends a character to your floor not found in wood from lumber yards today. Some salvage companies even re-mill the old wood, restoring it to its original, unblemished patina. Unmilled, this old wood retains the history of its lifetime, providing you with instant antiquity.

Not all salvage wood is antique. Wood reclaimed from old water tanks, wine barrels, or storage tanks has the same unique character, and even some of the darker tones found in old-growth wood, at slightly more than half the cost. Whichever you choose, or can afford, you will know you have done your part to keep old wood the treasure it should be.

When searching, either in the Yellow Pages or on the Internet, look for "reclaimed" or "salvaged" rather than "recycled" wood, if you want real wood as opposed to wood scraps glued together into a toxic burden neither you, nor your family, will enjoy.

Paints and Stains: Oil versus Latex

Americans use 3 million gallons of paint every day, or more than a billion gallons in a year. The paints and stains we use often are hazardous to us, our families, our pets, and our environment. Oil-based paints contain volatile organic compounds that vaporize at room temperature and make people sick. Some of the symptoms include headache, dizziness, shortness of breath, confusion, rapid heartbeat, and nausea. Latex paints also contain volatile organic compounds, which accounts for their odor, but to much less of an extent. Oil-based paints also are flammable, and when burned, release even more volatile organic compounds, because the primary catalyst for volatile organic chemical release is heat. Old paint, or marine paint, also may contain lead, polychlorinated biphenyls, cadmium, chromium, or mercury, all of which are lethal to humans, animals, and the environment. Although the Consumer Products Safety Commission banned the use of lead in consumer paints in 1978, some people still might have cans of old paint that contain lead. If you save old paint, sort your cans and dispose of old paint at your nearest recycling facility.

The use of oil-based paints during hot, sunny weather can contribute to the problem of ground-level ozone. Ground-level ozone is formed by nitrogen oxide and volatile organic compounds, which react with heat and sunlight to form ozone. Nitrogen oxides come from burning fossil fuels in cars, power plants, and industry. Volatile organic compounds are solid or liquid chemical substances that evaporate readily in the air and create particulates that cause their distinctive odors. They can come from car exhaust, gasoline, solvents, and paints, to name but a few sources. The result is toxic air pollution, and your state or local government will issue ozone air alerts on days when this problem is especially prevalent or widespread.

Even at very low levels, ozone pollution can cause or exacerbate health problems such as allergies, asthma, pneumonia, and bronchitis. Repeated exposure over several months might reduce people's ability to breathe properly, by doing permanent damage to their lungs. Since most people spend more time outdoors in summer than in winter, populations at risk for this damage include children, seasonal workers, joggers, sun worshippers, and gardeners.

Ozone pollution does not just affect people and pets. It is the direct cause of $3 billion to $5 billion worth of crop losses every year, because ozone interferes with a plant's ability to photosynthesize nourishment, making it more susceptible to disease, insects, other forms of pollution, and extremes in weather. This injury is particularly apparent on the surfaces of leaves, where ozone poisoning forms small, circular lesions like blisters that turn dark purple to black and spread over the leaf in distinct and readily identifiable patterns. Plants especially susceptible to ozone include common milkweed, and scientists have begun to track and assess ozone pollution through its effect on these plants.

Put a few drops of water in the paint you are using. If the water does not blend into the paint when you stir it gently but remains on top, you have an oil-based paint. Oil, or solvent-based paints, which are formulated using compounds such as linseed oil, petroleum distillates, alcohol, ketone, ester, or glycol ether, usually have a word or phrase on the can label to identify them. These words and phrases include but are not limited to: oil-based, alkyd, urethane, epoxy, clean with mineral spirits or paint thinner, contains petroleum distillates, or combustible. These paints are toxic, and leftover paint always should be taken to a recycling drop-off point or facility for disposal.

If you must use oil-based paint, ventilate the room or area well and wear a mask or respirator. Wear long-sleeved shirts and pants; skin breathes too and can absorb an amazing variety of chemicals. When you are through painting, either store the paint by covering the can with a plastic, household film such as plastic wrap and then hammering the lid in place, or, if the can is empty, recycle it at your nearest metal recycling facility. A paint can is considered "empty" if you cannot get any paint from it with a brush or if it does not drip paint when turned upside down. In some states, you can put the empty can in the trash as long as you leave the lid off so the collector can verify the can is empty, but check with your state or municipality to verify.

Do not put oil-based paints down the drain; the volatile organic compounds go into your drinking-water supply, which is already compromised. Do not put them down a storm drain either; they will contaminate lakes and streams or the ocean if you live near one. Also, do not allow the contents to dry into a cake, or lump, and then attempt to dispose of this in the trash, as this is illegal in most states. Most states have special collection points and recycling programs to dispose of oil-based paints and stains.

When you are through painting with an oil-based paint, you will have to clean your brushes and your skin with solvents, such as those used in the manufacture of oil-based paints. You may use either paint thinner or mineral spirits, because these work fastest to dissolve paint residue. You cannot put these down the drain either or dispose of them in the trash. You can filter the tainted paint thinner with cheesecloth and reuse it, but you then have to find a way to dispose of the cheesecloth, which is now a highly flammable rag emitting high levels of volatile organic compounds into the air until it dries. You can store the cleaned thinner in a glass, not plastic, jar with a tight-fitting lid to prevent additional volatile organic compounds in your air. Place it beyond the reach of children, who might view the contents as lime or cherry Kool-Aid, and in a location where it will not accidentally be tipped over and break. Alternatively, choose latex paint. Latex, ingested, will make people vomit. Paint thinner and minerals spirits will kill.

Paints are manufactured with a high level of volatile organic compounds so they will dry faster. Latex is better than oil-based paint, and a paint that has a flat finish will release fewer volatile organic compounds than a glossy paint. You can do your part to prevent pollution and ozone by buying only as much paint as you need. To do this, measure the walls in the room, or rooms, you plant to paint, and consult your local paint dealer, who will help you calculate the amount of paint needed to complete the project.

Choose latex paint where possible or acrylic paint. Modern latex and acrylic paints are nearly as scuff-proof and durable as oil paints. Latex paints have been formulated to work well in damp areas such as bathrooms, in high-traffic areas such as kitchens, on block walls in basements, even on the outside of your home, and to last as long as oil-based paints. In addition, when it comes time to scrape your siding, you will not be adding toxic paint chips and dust to the soil, air, and your lungs. Latex paints usually clean up with water, or soap and water, reducing the amount of volatile organic compounds you are adding to the air and the environment, while oil paints must be cleaned up with paint thinners or similar toxic liquids. Latex paints are commonly vinyl or acrylic compounds that dissolve readily and clean up quickly. Empty cans can go in the trash with the lids off. You can get rid of leftover paint by mixing it with kitty litter or newspaper shreds and letting it dry, then throwing the resulting dry mixture in the trash as well. Or you can store the paint for touch-ups. Just be sure to cover the can opening with plastic wrap before putting the lid back on, or the paint will dry in the can.

Stains

Wood stains are based on either pigments or dyes. Pigments are super-fine particles of inert compounds, frequently clay, which are used to make red or yellow ochre. Titanium dioxide, another natural compound, makes a brilliant white paint or paint base. Cadmium makes pigments in the green to orange spectrum. Copper makes pure green, and indigo, a plant, can be dried and pulverized to make blue.

Dyes come from plants. Indigo, mentioned above, performs as both a pigment and a dye. Other dyes include henna, cochineal bugs, and alkanet, all of which make varying shades of red. Dyes are water soluble. The most common type of dye stain comes in powder form and is meant to be dissolved in water to produce the desired color concentration. Water-soluble dye stain powders are environmentally safe and add color to wood. Unlike pigments, dyes penetrate wood, so the homeowner can achieve much better penetration of the wood without obscuring the grain.

Pigments and dyes also can be synthesized in a laboratory. Many modern pigments and dyes are synthetic, and although not toxic themselves, may incorporate toxic substances in their remanufacture into paint. The main difference between pigments and dyes has to do with the depth of the color they produce and the degree to which they obscure the grain pattern of the wood. Contrary to common conception, pigments do not penetrate the wood. Dyes, on the other hand, are incorporated into the cellular structure of the wood. Because of this, dyes tend to produce more transparent, natural-looking results.

Stains are characterized by the type of solvent used in their production. Stains are oil, water, and sometimes alcohol based. Each of these solvents affects the way the stain looks and handles. Unfortunately, oil stain is still the most common type of stain, leading water four to one. Oil-based wood stains contain pigments as opposed to dyes. They are high in volatile organic compounds, hence the smell. Oil stains employ mineral spirits, or petroleum distillates, as the solvent base. Mineral spirits are very flammable, volatile organic compounds. Linseed oil is another component of oil stains.

It often is treated with special acids so it does not soak too deeply into the wood. Oil stains also contain a thickening agent, often proprietary. There are two types of oil stains. These are penetrating oil stain, which sometimes bleeds and fades, and wiping oil stain, sometimes called pigmented stain, which is more consistent and does not streak.

Pigments do not penetrate the wood but instead collect in microscopic crevices in the surface of the wood. Very hard woods, such as maple, take pigmented stains poorly. To get a pigmented stain to stick, you need to create surface irregularities. A good method is sanding, using nothing finer than 100-grit sandpaper. Additionally, pigments are insoluble and can cloud the grain if the stain is not mixed frequently.

Water-soluble dye stains penetrate wood and are much more environmentally friendly during use and disposal. They do, however, have a few drawbacks. They might raise the grain on wood, and they are subject to fading. The color-fastness of dye stains has improved considerably over the past five years, with the development of non-grain raising (NGR) dyes, which come pre-blended in an alcohol base with a chemical added to slow the drying process. Even so, these dyes will still fade over time, while pigments are permanently colorfast. Concentrated non-grain raising dyes, such as Homestead TransTint Dyes, can be mixed with water or denatured alcohol to produce water-based wood stains and have remarkable non-fade properties. Because alcohol dries almost instantly, this dye does not allow much time or opportunity to correct mistakes, and you should practice before you begin, on a scrap piece of wood.

Water-based stains require some sanding after they are applied. This can be a significant problem if you are refinishing all the doors and trim in your house. If your wood is pine, which takes stain erratically, particularly water-based stains, you might have to pre-seal all your wood before you stain and sand after. This might seem like too much extra work, but the environmental benefits to your family will pay off over time.

Water-based stains are excellent for keeping that antique look in salvaged furniture. New paint hides the distressing that antique lovers want to preserve in their furniture, but water-based stains, applied with a sponge rather than a brush, will give a clean, finished look without hiding the decades of wear and tear you cherish.

All stains should be sealed to protect it from moisture, messy fingers, and dirt. You can use water-based sealers on water-based stains, but you cannot use varnish, and you should not. Many companies boast that their sealers are water based, when they are actually reformulations of old ingredients. Lacquer, for instance, may be advertised as acrylic based, and acrylics are made from wood or cotton. However, lacquer also contains xylene or toluene as the vehicle for the nitrocellulose cotton or wood fibers. Shellac is made from the resinous secretion of the insect Kerria lacca, mixed with ethanol. Ethanol can cause eye irritation and respiratory system difficulties. Tung oil also is derived from nature but is polymerized, or chemically treated, to make it usable. Varnish, made with acrylics or urethanes, also contains petroleum-derived and synthetic solvents and emits the most volatile organic compounds of any finish. Polyurethane finishes are plastics, and their production produces dioxin. Many water-based sealers are acrylic or urethane-based. Urethanes are compounded of isocyanates, the same ingredient used in pesticides. Many water-based sealers advertised as eco-friendly contain glycol ethers, a component in antifreeze.

A few good water-based sealants are made from plants, such as soy. These are called bio-based sealers. One is SoyGuard, made from the simple soybean. It combines durability with safety, has little odor, and is mildew resistant. There also is the potential for a corn-based sealer, based on polyactic acid, which is currently being used to make plastics but is still in the development phase as a wood sealer. Water-based wood sealers have a single disadvantage — they take a long time to dry.

Spray Paints

Spray paints involve the usual oil-based paint substances plus an atomizing effect from the aerosol can. Invented in 1949, aerosol cans have been linked to thinning of the ozone layer. Chlorofluorocarbons (CFCs) were banned in 1994, except for use in inhalation aerosols for treating asthma. Modern aerosols use liquefied petroleum gases (LPGs) such as propane, isobutene, and n-butane. Dimethyl ether is used in air fresheners and some spray-on deodorants and feminine products. Gases such as compressed air and nitrogen, both non-soluble, are recently finding a greater share of the market, as is simple carbon dioxide.

Liquefied petroleum gases, such as isobutene, are asphyxiant gases. That is, they displace oxygen in the body below levels needed to sustain life. The first symptom of overexposure is a feeling of not getting enough air. After that, users experience dizziness or drowsiness, a dulling of the senses, impaired judgment, nausea, and vomiting. If exposure is continued, unconsciousness, coma, and even death can ensue. Long-term exposure to liquefied petroleum gases can cause nosebleeds, rhinitis, oral and nasal ulcerations, conjunctivitis, bloodshot eyes, anorexia, weight loss, lethargy, fatigue, and damage to the central nervous system.

Dimethyl ether is a common propellant in household aerosols. In studies done on rats, changes in the blood were reported and involved bone marrow, the spleen, and the thymus gland; in two tests, the thymus gland atrophied. In one test, the number of white blood cells was more than doubled. Damage to male sperm and female eggs occurred at all concentrations and increased over time with repeated exposure. The results were a reduced number of conceptions, pregnancies, and live births, and newborn rats tended to exhibit skeletal malformations.

Nitrogen and carbon dioxide are naturally occurring gases. Compressed, they can form the basis for an aerosol can. However, if all spray cans made today were converted to nitrogen, our atmosphere, and eventually our waters, would all be dead zones like the growing patch in the Gulf of Mexico. Nitrogen levels in Tampa Bay are already four times higher than they were a mere 50 years ago. Nitrogen can cause higher concentrations of ground-level ozone; fall as acid rain, damaging man-made structures and acidifying soils; or deprive bodies of water of oxygen, resulting in eutrophication, or the dead-Gulf effect.

Paint Sprayers

Carbon dioxide already has been linked to global warming, so little more needs be said about this form of delivery in aerosol cans. The best recourse for those of us concerned with our health and the health of the earth we inhabit is to use pump sprayers, as opposed to aerosols. A can of spray paint might be cheap and convenient, but a single spray painter, which usually comes with its own air compressor, will cost about $80 and save you time, money, and paint over the long run. You can find these spray painters on the Web and in some hardware stores. Wagner, the first to patent a product, reportedly still makes the best and cheapest compressor-driven spray painters. Sharpe and Devilbiss both make high-volume, low-pressure (HVLP) spray guns, which are even more environmentally friendly and just as inexpensive as compressor-reliant spray painters. They require a bit of practice but deliver an even coat that dries quickly so there is no dust smudging. Once you get the hang of them, you can try adding color between finish coats to bring depth and highlights to your project. Unlike compressor-driven sprayers, they do not lay down a noxious fog of overspray, so cleaning up afterward is easier and you have not wasted as much paint.

Water-based Spray Paints

If you cannot afford a spray gun, there are water-based spray paints on the market. Krylon H2O is nonflammable and nontoxic, comes in a variety of colors, and cleans up fairly easily with soap and water. It costs about the same as a toxic can of spray paint, or $4, and smells much nicer. The "contents under pressure" on the label probably indicates a naturally occurring pressurized gas like nitrogen oxide or carbon dioxide. Krylon H2O qualifies under The California Clean Air Act, and California is picky about pollutants. You can find it at Ace Hardware. Durability is as yet untested and unrated; the paint is too new on the market to have generated reviews, but I used it and found it as initially scratch proof as any toxic spray paint I had used.

Rust-Oleum also has a water-based paint, called Aqua. It comes in eight colors, and the acrylic base is made from cotton or wood in the form of nitrocellulose fibers. Competitively priced, Aqua is available wherever Rust-Oleum paints are sold. The colors are not spectacularly different, mostly in the red and blue range, but Aqua is as easy to clean up as H2O and smells just as good.

Auto-Air Colors are water-based paints used primarily in vehicle painting and detailing. These are also acrylic and release only water vapor, not volatile organic compounds, when used in spray guns. Nor are reducers, hardeners, or setting agents required. The result is auto painting freed of time restrictions and product limitations. Auto-Air Colors are non-reactive, meaning the new paint will not be affected by the release of trapped solvents in older paint layers, and these paints need only half the volume of urethane to complete the same job. No primer is required, and Auto-Air can be applied directly to metal, aluminum, fiberglass, urethane plastic, and other common surfaces. Paint left in the spray gun will remain viable for months, making this paint ideal for home workshop painters, and it comes in 200 colors, including exotic finishes such as metal-or-pearl flake or iridescents.

Toilets

Toilets are the biggest users of water inside homes, using almost 27 percent of a household's water consumption, which is 6 percent more than clothes washing and 10 percent more than showering.

You can verify this yourself by taking a shower with the drain blocked, providing you have a big-enough drain pan, and measuring the volume of water in the pan when you are through. My shower, a standard 3-by-3-foot model, had 2.5 inches of water in it after five minutes. That is roughly two cubic feet of water. You can fill your washing machine with water and measure that as well. It will be about six cubic feet. A toilet tank, roughly 2 feet long and 6 inches wide, fills about a foot deep, or slightly more than a cubic foot of water, but it gets flushed at least ten times a day. If it is large, or old and leaky, it is using even more.

You can replace your old, leaky toilet with a low-water toilet, a dual-flush model, or even a composting toilet if you have a septic system. There are even toilets with a sink in the lid and one-piece toilets.

Low-water toilets

Replacing conventional toilets with low-flow models has the potential to save up to 12,000 gallons of water per year in a typical household. You can choose from the more common gravity-fed tank or a pressure tank. The first relies on the water pressure coming in to your home, as well as the pressure maintained by your plumbing, and requires 15 pounds per square inch of pressure to work properly.

Ninety-nine percent of residential toilets are gravity fed. Gravity-fed toilets work on a simple principle of displacement. There are three water reservoirs: the tank, the bowl, and the sewer. Water flows from the tank to the bowl and finally into the siphon tube, a U-shaped reservoir behind the bowl that rises, then descends, to the sewer pipe. This tube is always partially filled with water to prevent sewer gases venting into the room. When you flush, water is transferred from the bowl to the siphon tube, and this continues until the bowl is empty of wastes. At that point, the siphon stops working because air pressure enters the tube, making that gurgling sound we are all acquainted with, and water again fills the bowl. When the reservoir is refilled, the toilet is ready for operation. The rate of displacement is entirely dependent on water pressure, and some of this pressure is generated by the amount of water in the topmost reservoir.

If you have city water, your local water resource will know the pounds per square inch (psi) of water delivered to your home. This, however, does not accurately reflect the water pressure inside your house, which is determined by its plumbing. If you have an older home and complicated, or retrofitted, plumbing, you might have to consult a plumber. Homes having one continuous supply pipe, rather than several "feeder" pipes, will experience low pressure when two or more water systems, such as a shower and the dishwasher, operate simultaneously. You can rectify this problem, at considerable expense, by calling a plumber. Alternatively, you can do it yourself, if you have the know-how.

Low-water, or low-flush, toilets use about 1.5 gallons of water per flush, as opposed to older models, which can use between 3 and 5 gallons of water. In January 1994, the Energy Policy Act required all new toilets produced for residential applications to operate by these new standards, or 1.5 gallons per flush. Three-gallon flush models continue to be manufactured but are applicable only in commercial buildings.

A low-water toilet costs between $125 and $250. The savings in water costs are about $50 per year, so a new toilet pays for itself in 3 to 5 years. Low-water toilets are inexpensive but not without problems: They require more flushes to dispose of solids in the bowl, and even after several flushes, residue might remain; they clog easily, and vigilance is required to prevent them from damaging floors and making a mess. Newer models are better, but more than a quarter of homeowners are still trying to resolve the problems of their low-water toilets, which can cost several hundred dollars worth of a plumber's time to fix; double that if the toilet overflows and ruins the floor and subfloor.

If you have well water, you also have a pump. You can check the rating on your pump by examining the label. Pumps have to produce adequate pressure, which is about 50 to 70 pounds per square inch for the average household, and this pressure rating is based on their horsepower rating. They are also rated on gallons per minute (GPM) and gallons per hour (GPH). The average American household needs 10 GPM minimum and a 1.5 horsepower pump to deliver water to two or more sources adequately, without the pump cycling, or "choking." A pump this size costs about $800. You can spend less, but water pressure will be substandard, and it might take a long time for your low-flow toilet to fill.

Instead of replacing the pump, you can choose a tip-bucket toilet. This relatively new technology uses a bucket, located at the top of the toilet tank, which fills with water when the lever is activated. The water from the bucket is "tipped" into the tank and subsequently drains into the bowl to clear wastes. Because of the height of the bucket, water pressure in the bowl is increased. These toilets do not have flappers — that little tongue of rubber inside the tank that sits over the water opening — so there is little repair. They are, however, unsightly, and you will still have poor water pressure in the rest of your house.

Pressure-assisted Toilets

Pressure-assisted toilets look very much like a gravity toilet but have a second tank inside the main tank, or reservoir, and this tank forms pressure based on the amount of water pressure in your home and your toilet. Here again, you will need adequate pressure to make it work properly. If you have adequate pressure, this toilet will literally "push" wastes out of the bowl, unlike a gravity toilet, which uses water to pull the wastes into the siphon tube.

These toilets require 25 to 35 psi of water pressure and cost about $350, compared to a gravity-assisted toilet's average price of $150. They do not have a flapper and are similar to flush-valve commercial toilets in this respect. They are also not as quiet as gravity-fed toilets, and that rush of water and air can be both loud and disconcerting to the first-time user. They are better than low-flow toilets in that wastes seldom become trapped in the siphon tube, and overflows are rare. The downside is you have to take the lid off the tank and hold one hand over the inner tank opening if wastes do become trapped in the tube.

Vacuum-assisted Toilets

Gravity- and pressure-assisted toilets push water into, and through, the siphon tube. Vacuum-assisted toilets "pull." They have two airtight plastic vessels inside the tank, called the flush-water tank and the vacuum tank, and these are connected by a tube to the air-filled portion of the siphon tube. When you flush a vacuum toilet, the air in the siphon tube is sealed by draining water. At the same time, the water draining from the vacuum tank creates a vacuum that is "sent" to the air space in the siphon tube. This mild vacuum, aided by the force of water, sucks wastes down into the drain. First introduced in 1995, these vacuum toilets cost roughly $250, use an average of 1.6 gallons per flush, are much quieter than pressure-assisted toilets, and perform well in household applications, with few overflows. Because they use the same flapper-and-valve technology as gravity toilets, you will not need three hands to clear an overflow on the rare occasion it does happen.

True vacuum-assisted toilets are primarily commercial in application and, while using significantly less water, do require a plumbing system to be set up with negative air pressure in the lines. This can be prohibitively expensive, but if you have the money, you can check out the toilets featured by Evac Environmental Solutions.

Dual-flush, Pump-assist, and Air-pressure Toilets

Dual-flush toilets allow the user to select between two volumes, depending on the contents of the bowl. The lower flush requires between 0.8 and

1.1 gallons, the higher, 1.6 gallons. This eliminates the need to flush twice, as encountered by most low-flow toilets. Popular in Europe and Australia, where water restrictions are tighter, these dual-flush models are beginning to catch on in the United States. You can choose from a standard gravity toilet or a pressure- or pump-assisted model, with a number of control options. The Caroma, from Australia — the first dual-flush toilet introduced in North America — has two separate flush buttons on the top of the tank, one a solid circle, the other circle half-filled to denote lower flush capacity. Other models provide a split-handle, or a handle that turns either up or down for low flushing, and the position of the handle can be easily reversed. The Athena Company in Oregon offers a retrofit for older, high-volume toilets, and this has the up-or-down handle as well.

Some toilets have a built-in pump for low-water-pressure homes. The pump, a submersible unit, is housed in a lower tank behind the bowl. These toilets offer low water usage and dual-flush abilities, but the cost — $700 to $1,100 — is quite high and requires an electrical outlet to operate, something not commonly found behind most bathroom toilets. Bearing in mind the rule that water and electricity do not mix, I would have to give a thumbs-down to this particular innovation.

An innovation worth investigating is the air-pressure toilet, first manufactured for boats, buses, and other situations where water is scarce. A California company, Microphor, first put this design forward almost 20 years ago and has recently re-engineered the design to serve residential properties as well. It uses just one quart of water and runs off an air compressor, which can be located in the basement or utility room. This compressor can serve several toilets, but the cost of each toilet is from $600 to $900, and the compressor itself can cost hundreds of dollars.

Composting Toilets

Water conservation projects have dramatically cut residential water use in some areas. The city of Seattle has, since 1990, cut water consumption by 24 percent, even though the population has increased by 11 percent. Even so, toilets still use water — low-flow toilets use about 7,500 gallons per year — and water is a decreasing resource best saved for drinking. One truly significant invention to come along in the last decade to address this issue is the composting toilet. Composting toilets, as their name suggests, abjure water and turn toilet contents into valuable soil amendments. The Chinese have been doing this for centuries, without toilets, and are none the worse for it. They call it "night soil." I call it an excellent idea; I would rather drink water than flush it down the drain.

Composting toilets are the paradigm for water efficiency in any green-building endeavor, particularly where water is a limited commodity. Composting toilets are toilet systems that treat human waste by composting and dehydration to produce a usable end-product that is a valuable soil additive. These toilets come in a variety of models, shapes, and designs and will fit any décor. They use little or no water and do not need to be connected to sewage systems. They come in either batch-system models or continual-process models. A batch system has one to four containers that rotate into the toilet area, are filled, and rotate away. These containers are sealed; when the cycle is complete, the first container is ready to be used. In the continual system, waste is moved downward to a reservoir, where it remains for up to a year or until harvested. Composting is accomplished using microorganisms, invertebrates such as worms, and dehydration and evaporation. I visited a composting-toilet home recently and was surprised by the lack of offensive odors, which was one side-effect I thought could not be avoided.

These toilets are available in the United States through about a dozen manufacturers, who will provide instructions on their use and on the proper construction of reservoirs, if needed. To learn more, go to: **compostingtoilet.org/compost_toilets_explained/what_is_a_composting_toilet/index.php**.

All toilets made in the United States must meet standards set by the American Society of Mechanical Engineers, or ASME, and the American National Standards Institute, or ANSI. If you are buying a high-tech toilet from an exclusive manufacturer, check to make sure it meets the ASME/ ANSI standards. An Adobe version of this standard is available at: **www.ronblank.com/images/tot15a.pdf.**

Remodelers switching to low-flow toilets might discover their plumbing is inadequate. Older residential plumbing, from 1900 to 1960, is designed for a higher flow rate, and new toilets may not evacuate enough water under enough pressure to keep the pipes clear. If you have had sewer problems in the past, installing a low-flow toilet may make them worse. The best you can do is monitor your house's sewer system and call a plumber when necessary. Plumbers charge about $100 an hour, with a $25 service-call fee, but this is still less than the cost of the water you have been flushing down the drain all year.

Lighting

There are a number of ways to improve the lighting in your home without resorting to artificial light. The first would be taking down heavy drapes, which block most of the natural light coming into your home. Not only do draperies block light and obscure the view, but also they attract and hold common household allergens, such as dust, mold, mildew, bacteria, and odors; remember, odors are particulate, too. Most drapes have to be dry-cleaned, which means once every two five years, and you are replacing allergens with dry-cleaning fluid, a truly toxic substance.

If you use drapes to ensure privacy, or to keep out the cold, consider investing in window film, which provides one-way viewing — inside to outside, but not vice versa — and can also help insulate your window glass. Good film, properly applied, provides absolute privacy and can reduce heating and cooling costs by at least 25 percent and your carbon footprint accordingly. Neither is it difficult to apply. For more information, reread the section on window films on page 132, or go to: **www.wilwaylumber.com/howto/insulation/howto103.htm**.

If, after installing window film, you still want something to cover that expanse of glass for aesthetic reasons, choose sheers. These come in dozens of sizes, styles, and colors, from elaborate, frothy-white kitchen window sheers to elegant single panels for a dining room. They are about a fourth the cost of draperies and provide the illusion of draperies, without the weight and light-blocking capability. Most also are washable, though they should not be dried on a heat setting, as this will cause them to wrinkle or shrink.

You also can choose light-filtering shades for a south or west window, though these will obscure more light than window sheers. These shades come in flat or pleated styles, on an overhead bracket such as Venetian blinds or on a roller. Some window manufacturers are incorporating light-filtering shades into their windows. Pella makes such a window, as does Andersen under the Eagle brand. Reviews indicate Andersen is the best.

If you do not want to cover the window but object to the stark look of the window frame itself, you could put up a short valance and two narrow panels, either inside or outside the frame. This effectively "dresses" the window without blocking your view or the light.

Conventional Skylights

Skylights allow nature, rather than your utility company, to light your home. This natural lighting, called "daylighting," can, if properly installed, save up to $50 per year in lighting costs. Conventional skylights consist of an acrylic dome on the roof, with a flat, acrylic panel at ceiling level. Older models provide between 100 and 200 watts of light.

Size and location are the two most important considerations when choosing a skylight. A skylight two feet wide will illuminate a room 20 feet wide.

More is not better; only so many lumens are needed to light a room, and larger skylights will let in more heat in summer and more cold in winter. The location of your skylight is equally important. If you live in Arizona, you will want a north-facing skylight, preferably one that has a built-in shading system; if you live in Minnesota, you will want to install your skylight on the south side of the roof. When considering location, do not ignore the trees that surround your home, as these can inadvertently shade skylights and make them little more than holes in your roof. Additionally, place skylights as close to the peak, or ridge, of the roof as practical. This will prevent rain and snow puddling or freezing and damaging the flashing around your skylight.

Flat skylights provide direct midday light and might leak. Domed skylights catch more light and diffuse it better. Buy skylights based on their energy-efficiency rating, and add insulation in the roof according to the manufacturer's recommendations. Make sure to use sufficient flashing and caulk all exposed edges carefully when you are through installing; leaking roofs and wet ceilings defray all the cost advantages of using natural sunlight to illuminate your home.

In addition to free lighting, skylights also offer a remedy for Seasonal Affective Disorder (SAD). SAD is a direct result of shorter days and less natural light in winter. It is most common among people who live in northern climates, where seasonal differences in daylight can be quite extreme. Standing under a skylight on a cold but sunny winter day can trigger chemicals in the brain, called serotonin, which make SAD sufferers feel livelier and happier. The Mayo Clinic (**www.mayoclinic.com**) has some good information on the subject.

Having a skylight that opens to admit fresh air can be a benefit on a hot summer day. Heat rises, and opening a skylight can reduce cooling costs in a room by 20 percent. However, this feature is fairly expensive and might also be less airtight than a sealed skylight. Invest the extra money only if you live in a climate free of torrential rains or massive snowfalls.

Solar Tracking and Tubular Skylights

If you want the best that technology has to offer, consider solar tracking or tubular skylights. The first, using highly reflective mirror panels to track sunlight, can provide up to four times more light, even on cloudy days, or in winter when the northern sun stays low on the horizon. Solar tracking skylights also function up to twice as well as conventional skylights during the morning and evening, when sunlight is less available. The panels need no electricity to operate and annually provide 42 percent more light gain during daylight hours than conventional skylights. They are quite expensive, at about $1,000 for a small unit, but worth the expense in cost savings on artificial lighting averaged over five years.

For additional light in dark areas, without artificial lighting, consider a tubular skylight. These revolutionary devices capture sunlight in a clear, multi-faceted acrylic dome and use refraction technology, or prisms, to redirect the light into the room via a diffuser lens. The illumination provided can be as much as 700 watts on a sunny day or as little as 100 watts on a cloudy day, but the cost to operate is zero, compared to a 100-watt light bulb, which uses 875 kilowatt-hours per year and costs about $90. Because the cost of the skylight is about $150, not including installation, you can pay for the skylight itself in two years or less.

The refraction technology of many tubular skylights amplifies the incoming light so that a narrow aperture in the roof, maybe about one foot in diameter, provides enough light to illuminate a room 10 feet wide and 15 feet long. The inner "eye" is removable for cleaning, and all models feature solid, leak-proof construction and corrosion-resistant exteriors. Some manufacturers also offer a lifetime warranty.

Ambient Lighting, Sources, and Effects

Ambient lighting is, simply, all the light in a given room from all sources, whether natural or artificial. This can include light from windows or skylights, electric lighting within the room either during the day or at night, lighting from television screens, computer monitors, liquid crystal displays (LCD), appliances, streetlights, or nearby commercial buildings. Ambient light represents not only a quantity of lumens, but also the way in which this light is distributed throughout the room. The term is used to separate this type of lighting from task lighting, such as a light fixture above the sink or a reading lamp in the living room. Although task lighting contributes to ambient lighting, it does so in such a way as to leave most of an area dark. Some people prefer these "islands" of light, and they are sometimes called mood lighting, even though mood lighting is use-specific rather than location-specific.

Light has a tremendous effect on our mood and sleep patterns. A recent study, performed by Dr. Philip D. Sloane of the University of North Carolina-Chapel Hill and his colleagues and published in the *Journal of the American Geriatrics Society*, suggests that bright lighting during the day improves sleep performance at night among older populations, particularly those afflicted with Alzheimer's. The reason is fairly simple: We are all equipped with a circadian rhythm, a phrase taken from Latin and meaning "about a day." When these rhythms are associated with a day/night cycle, they are known as diurnal rhythms, or a "diurnal clock." Light during the day signals a chemical in the hypothalamus area of our brains to keep us alert. Reduced light triggers other chemicals that make us sleepy. These rhythms also control appetite, metabolism, and even reproductive cycles; we are hungriest in the morning or early afternoon, most active during daylight and become pregnant most often in the spring, when the days start getting longer. A Harvard study has even concluded that babies born in the winter/spring — or those conceived around March — are both physically larger and smarter than babies born in summer. In 2002, scientists at the Max Planck Institute for Demographic Research in Germany concluded that people born in the winter or spring live longer and suffer fewer illnesses in old age.

We cannot change the day we were born, but we can improve our health simply by improving the lighting in our homes and places of work. Living areas, especially kitchens, should be brightly lit during the day, preferably with natural lighting from windows or skylights. You can also choose task lighting focused on work areas. Offices and dens, where mental work takes place, also need bright light to keep the mind moving but do not need the additional distraction of a view — either from a window or into another cubicle — and these are ideal areas for overhead fluorescents. Bedrooms and rooms used primarily in the evening for "unwinding" require softer lighting and benefit from window films or curtains that block outside light. In the presence of appropriate lighting, mood and appetite improve, and performance has been shown to increase a full 25 percent. If your home or workplace is poorly lighted, make the necessary changes or request your employer to do so; nothing except air and water is more important to human well-being than appropriate amounts of well-placed and well-directed light.

Energy Systems

This section will look at solar energy in various forms, including solar electric generation, wind generation of electricity, and geothermal options, including ground-source heat pumps, any of which can assist with a home's power needs or, if large enough, take a home "off the grid." Even on-grid homes can benefit from these resources by returning electricity to the grid for a financial payback from the utility company.

Solar

Heat and energy from the sun can be used in several forms. Passive solar relies on the sun's heat, and no mechanical assistance, to warm spaces. Hybrid systems combine passive solar and circulating fans. The sun's heat can be used to heat water, or via a solar or thermal blanket to heat pools.

Solar electric systems produce electricity from sunlight via cells made from light-absorbing monocrystalline silicon wafers. Using the photovoltaic (PV) effect, these semiconductor materials absorb sunlight, separate charged electrons, "carry" them on metallic contact plates, and transmit them through wiring, in a process known as the photoelectric effect.

A single cell produces a limited amount of electrical charge; joined with about 50 others in a panel, or module, these cells can produce from 10 to 300 watts. Joined panels, or arrays, can produce even more power. Ten to twenty panels can power a house.

A solar electric or photovoltaic system consists of cells made into panels, installed as an array using racks and interconnected wiring, with an inverter that transforms the raw, direct electrical charge to a 120- or 240-volt alternating current. A solar electric system might also include such items as an anti-static baffle to prevent electronic interference with televisions and computers and specially designed batteries to store excess electricity.

There are certain items to be taken under consideration before installing a solar electric system. They involve size, which includes both the size of the system and the size of your roof; the geographic location; the cost; and the system's ultimate use, such as, will the system provide only part of the electricity for the home, or all the electricity? Cost and size are directly related. The cost of a kilowatt-hour of electricity declines as the system grows larger.

A typical system designed to provide electricity for a home that uses about 800-1,000 kilowatt-hours of electricity per month consists of about 10 solar electric panels and requires between 200 and 400 square feet of roof space, depending on the panel's efficiency rating. Photovoltaic-panel efficiency ratings have improved greatly over the past ten years, but more efficient panels are also more expensive panels, because emerging technology is not cheap. Your house will reasonably have to be about 20 feet wide and 40 feet long to accommodate the system described above, since solar goes only on the side of the roof facing the sun. Your roof will also have to be in good condition and capable of bearing the weight of the panels, racks, wiring, and other accessories. Smaller systems, designed only to replace electricity used in lighting, are also available and can be upgraded over time by adding more panels.

The pitch, or angle, of your roof will also determine the effectiveness of your solar electric system. In northern latitudes, in winter, the sun is lower in the sky throughout the day, and less sunlight is absorbed. If you have a standard 5:12 roof, defined as 5 inches of rise per foot to the peak, the sunlight will strike the panels at an oblique, rather than direct, angle. If you install a fixed system on the roof, as opposed to a moveable system at ground level, you will not be able to take advantage of all the sunlight available. With today's solar technology, this might be less of a consideration than it was ten years ago, but it still affects solar collection. You also need to determine whether the sunlit side of your roof faces due south. The best orientation for solar collection in winter is an 8:12 pitch and due south. However, 45 degrees east or west of true south will not impair your solar electric system efficiency to such an extent you need to abandon the project. Other considerations might.

If your south-facing roof is obscured by trees, and the trees are on your neighbor's property, you might have to negotiate to get the trees trimmed or removed. Some states provide a "solar easement," allowing you to prevent neighbors or future developers from planting tall trees or building structures that would obscure your roof's sunlight, but these laws do not apply to existing buildings or trees, which are by law "grandfathered in." If you can convince your neighbor to have the trees trimmed or removed, you will have to pay the cost of doing so. In fact, some state and federally mandated rebate programs require your solar installation to be unshaded for a certain number of hours on a certain number of days during the year. Since these rebates represent a significant portion of the cost of solar installation, you should determine beforehand whether you can meet the standards. If you live in the shadow of a tall apartment building or condo, you might want to reconsider your investment. If, after researching your site's solar value, you are not certain, consult a professional. He or she will provide a solar site analysis that, although not free, will prevent you from spending money on a project that will not be beneficial for you. If, after an analysis, you are unable to use your roof for one of the above reasons, you can choose a ground-level photovoltaic array, either fixed or mounted on a tracking device that will follow the sun.

Solar electric systems use both direct and scattered sunlight to generate electricity, and most areas in the United States have ample sunlight to operate small solar electric systems. However, the return on investment will vary; cold, northern latitudes such as Chicago will provide less electricity per panel, particularly in the winter, than warm, southern areas such as Houston or Phoenix. Do not let this dissuade you if you live in Wausau, Wisconsin, however, as photovoltaic technology is now sufficiently advanced to capture even minimal sunlight.

The cost of a solar electric array is dependent on its ultimate use. Will you be off the grid — that is, not connected to a local utility's power supply lines — or grid-tied? If you are tied to the grid, using your photovoltaic array to provide electricity only during sunny days and seasons and allowing the utility to supplement your electrical needs, will you subscribe to net metering or have battery backup for your system to store excess electricity generated?

Let us suppose you are grid-tied and want to use your solar electric system to cover only the lighting requirements of your home. If you live in an existing home, your local utility company can help you determine how many kilowatt-hours you use for lighting, and this will determine the size of the system you need. If you are building a new home, you should consult an expert who will also install the system for you. Photovoltaic technology is sufficiently complex that you do not want to do it yourself.

Before you consider any system, give your house an energy audit and find out where you can conserve if necessary. For example, if you want whole-house solar electric but you experience cloudy days in the winter, consider washing dishes by hand and washing and drying clothes only on sunny days. If you have already done an energy audit and wrapped your hot water heater, install the system you need. You can always add panels later, and panel efficiency continues to improve yearly.

I have already explained most of the components of a solar electric system. Some, like wiring, are self-explanatory. The inverter, which is like the heart, is the second most important element of a solar electric system. Electrical current naturally flows as direct current. Direct current is a one-way trip; alternating current can go in either direction and flows gradually, in smooth voltage transitions, in what is called a "sine wave." Solar panels create direct current, which must be transformed to alternating current to be usable inside a home, or sent back along the grid to a utility provider. Inverters can be either true sine-wave inverters, which you must have if you plan to tie to the grid and use your utility's net-metering feature, or a modified sine-wave converter, which can be used to power many alternating current loads, though this latter feature does not provide the same quality of current your utility company will require if you plan to sell electricity back to them. There are also square-wave inverters, which produce electricity but not the peak voltage required to operate many electronic appliances such as computers, televisions, and DVD players.

Solar electric systems are grouped by their rated power output; that is, the maximum amount of power they produce when receiving 1,000 watts per square meter of sunlight at about 58°F. Any system rated between one and five kilowatts is adequate to the average home.

The capital cost for a solar electric system is significant, ranging from $8,500 for a small backup system to $13,000 for a small residential application of 2,000 watts and up to $45,000 for a 5,000-watt system. If you are a homeowner, you are paying about $2,000 per year for electricity. Electric costs continue to rise. In five or six years, with rebates, you will have paid for your system, and solar rebate programs abound. You may also get subsidies or tax incentives for installing solar, and these can come from the federal government or the state as a credit on your income tax return. They can also come from your utility company and sometimes even the manufacturer. For more information on rebates and tax incentives, visit: **www.affordable-solar.com/state.rebates.htm**.

The operating costs of a solar electric system are considerably smaller, but the time required to monitor its condition is greater than that of being connected to the grid, particularly if you use batteries for backup power. Heavy-duty, deep-cycle batteries, recommended for solar or wind-powered electric generation, require monitoring. Trojan makes an excellent 6-volt battery. Do not use car or marine batteries. Batteries wired in series can increase amperage, safety vents can act as water recyclers, and batteries must be kept in a secure, stable location. A battery monitor is also recommended. Because batteries can never be fully discharged without damaging them, you will need to purchase between 20 and 40 percent more battery backup power than you need, based on both your PV system's output and the amount of time you need to store power. If you are building off-grid, do not allow your solar contractor to "bid down" the cost or amount of battery backup to make the job cheaper; if you cannot use the power when you need it, you might as well not pay to generate it.

The final consideration with a solar electric system is obtaining permits. Before you put a down payment on a system, consult your local county, city, or municipal government. If you have a homeowner's association that regulates building and landscaping details, read the covenants. You may need both a building and an electrical permit and might require approval from your homeowner's association. Some state laws allow you to bypass homeowner's association rules, but get this in writing from the state before you proceed. Code requirements for solar electric applications vary, but most are based on article 690 of the National Electrical Code, which spells out the requirements for safe, reliable PV-system installation. This permitting process can be frustrating and time consuming, particularly if you are the first in your community to install such a system. Your local building department might not have developed regulations, and the permitting process will be a learning process. You can, of course, rely on your contractor to deal with the permitting details, but if you do, make sure you spell it out in legal documents who is responsible and who will pay to fix errors, if the permitting process falls through the cracks. For more information on permitting and codes, visit: **www.eere.energy.gov/consumer/your_home/electricity/index.cfm/mytopic=10690**.

Last but not least, choose a reputable contractor. Get at least three bids for the project, each clearly stating the maximum generating capacity of the system measured in watts and the total cost, including permits and taxes. If possible, have the contractor specify system capacity in alternating current, or the output of the system at the inverter. Request an estimate of the total annual output of the system in kilowatt-hours, realizing that sunlight varies from year to year, and no one can make a pinpoint prediction about efficiency. Then select the contractor you trust or the one who seems most knowledgeable. Ask whether the company has installed solar electric systems previously, and if so, where. Also ask whether you can consult with one of their customers, preferably one whose system has been online for a year. Ask to see the contractor's licenses, particularly an electrical license, and verify its validity with your state electrical board or local electrician's union. Find out how many years the company has been in business, and check it through the Better Business Bureau. Find out whether it offers a warranty, how long it extends, and what it covers. Finally, ask whether it also services what it sells, since you might encounter problems you cannot fix yourself.

If you are building a new home, there is a new roofing material coated with an electricity-producing film. This material works well in extreme climates and has the durability of more conventional products. Known as "thin film" photovoltaic technology, this coating can convert any surface into a solar-electric power source, and engineers are working to adapt it to house-siding technology as well.

Energy Trust of Oregon produced a Case Study on wind power initiatives, which you can read next.

CASE STUDY: ENERGY TRUST

Renewable Energy

Warren and Elizabeth Griffin

Wind and sun power Salem family's home

On a green hilltop in south Salem, the Griffin family is quietly leading the way to Oregon's energy future. With a wind turbine, solar photovoltaic (PV) panels and a solar water heating system, they are generating much of their own electricity and all of their hot water.

The 1.5-kilowatt wind turbine serves as a demonstration project for small wind systems in low-wind areas and is connected to the Portland General Electric power grid. It is expected to provide about 55 percent of the Griffin's electricity. With Energy Trust incentives and Oregon Residential Energy Tax Credits, the family was able to invest in both solar and wind power – and a more sustainable future for their children, Robby and Gillian.

"Our vision was to take a sustainable approach in building our home, and part of that is using solar and wind energy. In the long-run, fossil fuels aren't a viable option, so we want to help lead the way by using alternatives," said Elizabeth Griffin.

Goals

- Integrate renewable power into a sustainable home design
- Maximize electricity generation with clean, renewable wind and solar power
- Demonstrate feasibility of a small-scale wind system in a low-wind area
- Reduce reliance on fossil fuels

CASE STUDY: ENERGY TRUST

- Minimize environmental impacts

Strategies

With help from the Energy Trust trade allies Silverton Electric and Summers Solar Systems, the Griffins chose to install an 1800-watt pole-mounted photovoltaic system that tracks the sun and solar water heating system with an 80-gallon storage tank.

The Griffins applied for Energy Trust support for the wind turbine through the Open Solicitation program. Their building contractor Bilyeu Homes, installed the turbine.

Equipment Specifications

Solar Electric:

- Pole-mounted 1800-watt PV array with 12 Isofoton 150-watt modules
- SMA 1800-watt inverter

Solar Water Heating

- Solomax system with a 40-tube Termomax collector
- 80-gallon storage tank

Wind Turbine

- 1.5-kilowatt African Wind Power wind turbine on a 106-foot tower
- SMA Windy Bay 1800U inverter

Financial Analysis

The Griffins received cash incentives from Energy Trust and State of Oregon Residential Energy Tax Credits, which helped make the project more affordable.

Solar Electric

- $14,00 project cost
- $5,850 incentive from Energy Trust of Oregon
- $1,500 Oregon Residential Energy Tax Credit
- $192 estimated annual energy cost savings

Solar Water Heating

- $5,927 project cost

CASE STUDY: ENERGY TRUST

- $1,040 incentive from Energy Trust of Oregon
- $1,500 Oregon Residential Energy Tax Credit
- $192 estimated annual energy cost savings

Wind Turbine

- $23,208 project cost
- $10,500 incentive from Energy Trust of Oregon
- $1,500 Oregon Residential Energy Tax Credit
- $395 estimated annual energy cost savings

Project Benefits

- Reduced reliance on conventional power and fossil fuels
- Protection from rising power prices
- Combined estimated annual electricity generation of 7,939 kWh (unused power feeds into the PGE grid through a net metering agreement) and electricity saved is 2,600 kWh
- Avoided over 100 tons of carbon dioxide emissions

Project Team

Warren and Elizabeth Griffin

Energy Trust of Oregon

Silverton Electric (trade ally)

Summer's Solar Systems (trade ally)

Bilyeu Homes (home building contractor)

Portland General Electric (electric utility)

How can we help you?

Are you interested in having a solar electric or water heating system installed on your home? Energy Trust provides direct incentives to lower costs, supports a network of solar trade allies and provides standards to ensure quality.

CASE STUDY: ENERGY TRUST

Do you have an idea for a small wind, small hydro, biomass or geothermal project? Energy Trust may be able to provide funding for these, too.

To find out more, call 1-866-ENTRUST or visit our Web site at energytrust.org/SE.

Energy Trust of Oregon, Inc.

Wind

Generating energy from wind is as old as the Roman Empire. The earliest wind turbines were vertical axis machines, having sails mounted around the edges on vanes perpendicular to the ground and were used to grind corn and move water. Later, in Holland, fan-configured windmills were used to pump water and mill grain. These early windmills produced a maximum of 30 megawatts. When the United States first started to focus on alternative energy, around 1970, solar technology was scarcely advanced and extremely expensive; people who wanted to get "off the grid" turned to wind, a proven technology.

Though wind turbines now come in various shapes and sizes, wind turbines still resemble a standing fan, having a tower, and, usually, three long blades attached to a rotor on a horizontal axis. The tower, which can range from 80 feet to 120 feet or more for residential applications and more than 300 feet for commercial, or wind farm, applications, is necessary for two reasons. First, wind velocity increases with height. Second, a tower raises the turbine blades above wind baffles such as trees and land formations. Recommendations from experts suggest that a tower should be 30 feet above anything within a 300-foot radius, since air is very fluid, like water, and obstructions can generate significant amounts of turbulence, which will reduce the turbine's efficiency or even damage it.

Commercial towers are often solid objects; that is, a steel shell over internal parts. Residential towers are more commonly grid towers, or a collection of interconnected steel angles, or poles, sometimes anchored with guy-wires, which are metal cables fastened to the tower and anchored in the ground. A grid tower is more thermally effective because it allows damaging, high-velocity winds to pass through it.

The turbine itself, a series of fan blades attached to a central cam called a rotor, can be as wide as the turbine is tall, and this necessitates siting the wind turbine on a good-sized plot of land. Unlike solar, which can be placed on a roof and is virtually unobtrusive, wind power is highly visible and generates noise. Though small systems are no louder than a washing machine and will not interfere with television reception, they are still visually obtrusive and might make the permitting process more difficult. Those considering wind power should have at least an acre on which to locate the turbine. Even with these restrictions, wind power is an excellent alternative for those who live in areas where wind speeds meet or exceed 10 miles per hour and where electricity costs equal $.10 per kilowatt-hour.

Having a wind turbine can lower electric bills by 50 to 90 percent. The cost of a residential wind turbine is between $5,000 and $20,000, depending on size and service agreements; that is, whether you can perform routine maintenance yourself or want the installer to maintain the equipment. Your home will not have to be rewired to accommodate a wind power system, because it, like solar, uses an inverter to convert direct current to alternating current, and you can either use battery backup or feed the energy into your local utility's grid.

Configurations for wind systems have evolved dramatically in the past ten years. You can still buy the typical tower/turbine blade configuration, and these second-generation wind systems feature increased capacity and durability at reduced size, though tower height is still dependent on the terrain. The 1.8 kW Skystream 3.7 produces about 400 kilowatt-hours per month, stands from 34 to 70 feet tall, has a 5-year warranty, and costs about $9,000. The 10 kW grid-tied rating Bergey BWC Excel costs about $20,000, not including a tower, and can provide all the power needed for a remote, all-electric home in areas where wind power is rated average. Where there is no grid and batteries are used to store power, wind power systems cost about $4,000 per kilowatt. A typical 110-volt, 220 amp-hour battery storage system, including a charge controller, costs a minimum of $2,000.

Operating expenses, once the system is up and running, are about 2 percent of the initial cost, or, in the case of the Skystream, about $200 per year. Cost recovery on the initial investment generally occurs in 20 years, but this figure does not take into account increasing energy prices, interest on any loans, the accrued benefit to the property at resale, any state or federal tax incentives, or the actual resale cost of the turbine. It also does not take into account the cost of connecting to the grid, which in remote locations may equal half the cost of the system itself, or as much as $30,000 in more remote areas.

Once upon a time, your regional utility would install up to 500 feet of grid-connection line at no cost. This is changing. Smart building codes and efforts to prevent urban sprawl have caused states such as New Jersey to adopt new utility codes in which homeowners are required to foot the bill for connection. You may recoup these costs over the years as more and more houses are built in your area and additional homes tie into the grid, but the up-front costs are still yours if you are the first settler in an area. Before you consider a wind-power system, consult your local utility and factor in the cost of grid-tying before you decide how much you want to spend.

In spite of their appearance, wind turbines are very safe. You are 100 times more likely to be struck by, or have your house damaged by, a falling tree than a falling wind turbine. In spite of that, people still imagine all sorts of dire scenarios, because they do not know that wind systems shut down automatically in the event of a power outage and will not energize a fallen power line. Turbines also shut down in the presence of dangerously high winds because they have an overload switch. Lattice towers allow high wind to pass through, and icing slows the blades so that the ice falls rather than being thrown. A well-constructed, guyed lattice tower is no more of a hazard to curious children, deer hunters, and adventurous teenagers than is a water tower or radio tower and should in any case be fenced or otherwise shielded to prevent mishaps.

Vertical axis turbines, based on second-century design, have blades that are perpendicular to the ground. One of these, the Mag-Wind Vertical Axis Turbine, or MW1100, is a "magnetically levitated, axial flux alternator with programmable variable coil resistance," which, according to Lloyd Alter, writing in *Science & Technology* in the January 2007 issue, solves many of the problems previously associated with vertical axis wind turbines. Designed by Thomas Priest-Brown and Jim Rowan of Canada, it has a center hub that floats on a magnet, with coils at the outer ring having magnetized tips that generate power when moving, especially at high speed. Designed for residential roof applications, it is only four feet high and is reportedly almost inaudible when operating. More important, high or gusty winds do not cause it to shut down or malfunction, and when placed on steeper roofs, for example an 8:12 or even 12:12 pitch, it will output 100 percent more power than it does on a typical 6:12 roof, because it takes advantage of something called the "roof effect." The MW1100 can produce about 1,000 kilowatt-hours per month in the Midwest, where wind speeds average 10 miles per hour, and begins to produce energy at wind speeds of less than 5 miles per hour — which is very close to the 3.4 mile-per-hour cut-in speed of a tower wind turbine. The MW1100 also performs at a top speed of 110 miles per hour, which is very close to a standard turbine's 120 mile-per-hour maximum. At $150 per month for electricity, the MW1100 pays for itself in six years and was available beginning in February 2007.

Another vertical-axis wind turbine manufacturer is WindHarvest (**www.windharvest.com/turbines**). The Windstar 3000, which looks something like a mixing blade standing on end, stands 50 feet high and 75 feet wide. It works well in areas where wind speeds are below 16 miles per hour and uses standard generators, gearboxes, bearings, and other components available from a wide variety of vendors, so parts replacement is simplified. It is highly efficient in areas where the wind blows an average of 14 miles per hour and, because of its height, can be installed under taller, existing wind towers. It, and other WindHarvest turbines, can be fabricated in any ISO 9000 certified factory. Because these vertical axis turbines are shorter than the typical tower/turbine configuration, construction and installation costs are significantly reduced. In fact, these turbines are a do-it-yourselfer's dream, because the parts are easily assembled and do not require a great deal of expertise. The foundations are typically a 5-foot cube, and the parts can be transported in standard, 40-foot cargo containers by truck. The essential moving parts, which include the generator, transmission, belts, and braking mechanism, are at ground level, so maintenance can be done without climbing the structure and without a hoist. Vane repairs are seldom needed becausethe vanes are extruded from aircraft-grade aluminum, but when they are, the structure can easily be climbed. These wind towers are not as harmful to birds and bats as standard wind turbines because their spinning vanes are seen as a solid object. Most of WindHarvest's wind turbines are designed for commercial applications, producing 100,000 kilowatt-hours per year, but they do manufacture residential turbines at significantly smaller sizes and power production.

There are a number of vertical-axis wind turbine manufacturers who produce a product for residential use, and you can find them online. You will also see some surprising prototypes for newer, more efficient "eggbeater" wind turbines. These might look fascinating and make some amazing claims, but like all prototypes, they require years of testing to validate efficiency and operability. When choosing a wind turbine, buy from a manufacturer whose product has been on the market at least five years, and ask for customer testimonials. Always read the warranty.

Christine and Sam Simonetta, who live in Upper Michigan where winters are harsh, added a wind turbine to their Five-Star Home building project, and the information on that project is provided below.

CASE STUDY: CHRISTINE & SAM SIMONETTA.

Courtesy Sam Simonetta,

Lean, Clean Energy Services

Patrick Hudson,

Michigan Energy Office, Dept. of Labor & Economic Growth

Upper Michigan's Five Star Grant Home

Courtesy of Sam Simonetta, Lean, Clean Energy Services

Christine and Sam Simonetta along with their licensed builder Stephen M. Smith created a unique Five Star Grant Home. Their project incorporated a balance between energy efficiency, renewable energy, comfort, and beauty. It was the first Five Star home in the grant program's history to utilize wind energy. While the house is tied to grid power, Sam and Christine's wind turbine produces 80% (on average) of their electricity. In some months, it produces over 100%.

For starters, Sam's smart home design economized building materials, as he "stacked" the 1700 sq ft story-and-a-half house on top of the garage. He also engaged a saltbox design elongating the south-facing wall to incorporate passive solar. A wise design is where energy efficiency begins.

The Upper Michigan location and harsh winters demanded extra attention to insulation. The couple chose 6" Structural Insulated Panel walls that boast an insulating value of R-24, and their thermal siding takes the total wall value to R-27. The roof is insulated with 10" of Icynene, an innovative, non-toxic product that also doesn't promote mold growth, which will boost their roof insulation to an R-55. Next is a 92% efficient boiler for their radiant in-floor heat. To keep the heat in during cold months, they have triple paned windows that have a total unit value of R-4.3.

CASE STUDY: CHRISTINE & SAM SIMONETTA.

From the beginning, Sam and Christine aimed to balance aesthetics with efficiency. The home includes rock knobs and wood railings found on the property. They brought character into the new structure by using vintage doors and a classic, old window for interior day lighting. They even extended the reused category to include an old but sound Mackinac Bridge beam as the main support in the basement.

Other energy efficient features include Energy Star appliances, compact fluorescent bulbs, phantom load reduction including switches for any continuous energy drains, thermal mass exemplified by their soapstone masonry heater, and an energy recovery ventilator to bring fresh air into this tight house with minimum energy loss. To conserve water, they installed sink/shower flow restrictors and dual flush kits on their toilets.

With an eye to clean, renewable energy, they also wanted to keep the inside of the house as environmentally friendly and non-toxic as possible. They chose to showcase bamboo flooring in two rooms, as bamboo grows quickly and can be harvested every 3-5 years, and it is lovely, durable flooring. They've chosen low or no VOC paints and varnishes as well to minimize out assign of toxic chemicals.

To them, the crowning jewel is their 2.5 kW Abundant Renewable Energy (ARE) wind turbine that proudly stands atop Onota Hill overlooking Lake Superior. The turbine gathers many winds and produces clean electricity for this simply efficient Five Star home.

For further information pertaining to this Energy Project or similar projects contact:

Http//wwww.northernoptions.org/up5starhome/

Feel free to contact us if you have any ideas for case studies or other questions:

Michigan Energy Office, Dept. of Labor & economic Growth

P.O. Box 30221, Lansing, MI 48909

Phone: 517-241-6228

Fax: 517-241-6229

Or Tim Shireman at tashire@michigan.gov

(After I obtained this study, Sam wrote: "At the time we wrote that, we did not have much wind turbine output data to go on. We now know that the turbine is producing more than 100 percent of our electric energy needs, not the percentage indicated. Our average monthly usage is about 225 kWh and average production is closer to 300 kWh.")

CASE STUDY: CHRISTINE & SAM SIMONETTA.

Case Study Courtesy the Rural Renewable Energy Alliance, Minnesota and Joel Haskard, Co-Coordinator, Clean Energy Resource Teams (CERTs) Regional Sustainable Development Partnerships

Geothermal

Geothermal energy is heat derived from steam from heated, underground water that passes through a steam turbine and is converted to usable electricity. Some commercial geothermal systems use existing steam; others pour water into hot-rock fissures in the earth to create steam. The mantle of the earth provides the heat, underground water reservoirs provide the steam, and extraction is relatively straightforward and pollution free. However, since some geothermal sites tend to go "cold" and it might take 100 years for the earth to produce enough heat to warm the site again, there is debate whether geothermal is truly a renewable resource. The other side of the debate, that geothermal may overly cool the earth's mantle, is like suggesting that Yellowstone may cause the earth to grow cold.

In residential terms, a ground-source heat pump (GHP or GSHP) is a geothermal system, since few homeowners have the capital or expertise to drill 5,000 feet into the earth to find superheated water that can be "flash" converted into steam and fed into a steam turbine.

GHPs, also known as geo-exchange, earth-coupled, ground-source, or water-source heat pumps, have been around since the middle of the last century. They use the temperature under the ground as an exchange medium, in a process called transference, and work because the temperature four feet below the surface is a constant 50° to 55°F year-round, even in the hottest or coldest climates. Their efficiency, compared to air-source heating and cooling, such as forced-air furnaces and compressors, is 100 to 300 percent greater.

To make a GHP, coiled or curved loops of open piping, called the heat exchanger, are buried underground in a grid pattern. These pipes, usually of plastic, are filled with fluid, which is heated or cooled by the earth's temperature. The heated or cooled fluid is then passed through a pump and distributed to the rooms in your house.

The heat exchanger itself can be made up of horizontal loops, if there is adequate space without hard, rocky outcrops or underground shelves of rock. A heat exchanger looks very much like a grid inside a toaster that is laid flat on a counter. Where space is limited, the heat exchanger pipes can be installed in vertical loops — that is, facing up and down instead of flat. Vertical loops also overcome the problem of rocky soil. Another increasingly popular method is installing "slinky coils" of piping horizontally in a trench, providing more surface area for heat transference in a smaller space.

The second vital element of a GHP is the heat transfer fluid. Contractors in most areas of the country use a substance similar to antifreeze, with corrosion-resistant properties and a freezing point of more than ten degrees below the minimum requirement for both the system and the climate. These glycolic, or alcoholic, solutions are biodegradable and nontoxic and can minimize the amount of pumping power required for a given system. In warmer climates, where the heat pump's exchanger refrigerant temperature does not fall below freezing, water can be used as a circulating fluid.

Most residential GHPs are closed-loop systems. The pump is located in one room, and heated or cooled fluid or air is ducted or circulated to other rooms. GHPs are also able to supply a residence with hot water. Models equipped with two-speed compressors and variable fans can extend both the comfort range of a house and its energy savings. Compared to air-source units, GHPs are quieter, last longer, need less maintenance, and are not affected by the temperature of the air.

There are GHPs for any size residence, located in any part of the country, and they can be installed on almost any size lot, even under existing lawns, driveways, landscaped areas, or the house itself. If you already have a forced-air system, a GHP can use the existing ductwork with a few minor modifications, and in many areas, special financing is available to defray the cost of the system. The Department of Energy and the EPA support GHPs as energy efficient and environmentally friendly.

GHPs can save up to 50 percent of your water heating bill, are about the same size as a traditional heating/cooling unit, and the heat exchanger carries up to a 50-year warranty. Unlike traditional systems, there is no exposed equipment indoors or out and no noisy fans or compressors. Neither are there any combustible materials involved.

Air-Source Heat Pumps

Electric air-source heat pumps (ASHPs), though not truly geothermal, do use the temperature differences between outdoor and indoor air to cool and heat your home. Most often used in milder climates, ASHPs work because, in cold weather, the outside air is still warmer than the refrigerant in the system, which causes the refrigerant to boil into a gas. When this gas is compressed by your system, it can reach 120°F or higher, and it transfers the heat to the inside of your home. When temperatures outside are high, the refrigerant acts as a coolant, reducing temperatures inside your home.

High-efficiency ASHPs with the Energy Star label use less energy than conventional models. They also come with higher heating and seasonal performance factor (HSPF) ratings. These ratings are another measure of the heating efficiency of heat pumps.

ASHPs consist of three main parts: an indoor heat exchanger, an outdoor heat-absorbing unit, and a water storage tank. ASHPs cost about 60 percent less than traditional heating and cooling units and are much smaller than GHPs. They can be installed in the same space once used by an old boiler and require no regular maintenance.

ASHPs are somewhat less efficient than GHPs, but they do not require the extensive in-ground installation of GHPs, so installation cost and lot-size considerations are overridden.

Tankless Water Heaters

Tankless, or on-demand, water heaters are common in Europe, where energy prices are high. They did not arrive on the American scene until about 1985, when rising energy costs made alternative forms of heating water both essential and saleable. Tankless water heaters heat water only when it is required, as opposed to hot water tanks, which heat water all the time, using unnecessary amounts of gas or electricity.

Tankless water heaters are expensive in comparison to traditional water heaters. A tank heater costs about $600 and provides hot water to an entire house. A tankless heater providing enough water for a bathroom can cost $200, with whole-house models costing $1,000 or more, not including installation, which can be complex.

Tankless hot water heaters come in both gas and electric models. Gas models are efficient, unless they have a "standing" pilot light, which reduces their efficiency. Large electric models may require more electricity than your house can provide; smaller electric models still require rerouting your house's electrical circuits to each unit.

To decide what size tankless heater you need, determine the flow rates of the specific taps or faucets you want to connect, the temperature of the incoming water — which will vary significantly from summer to winter in some climates — and the number of taps or faucets that may be used at any one time. If you undersize the system, you might as well stay with a tank water heater. No one wants to bathe or wash dishes in cold water.

On average, a small tankless water heater will supply enough hot water for a bathroom and a kitchen with a dishwasher. This, however, depends on the number of people accessing the hot water and the time of use. You can also "zone" your tankless heaters by room. If a bath and laundry room are adjacent, you can zone your tankless heater to work for both. Your contractor or electrician can help you size your tankless heater for greatest efficiency and optimum performance. Some manufacturers will also respond to sizing questions in a timely manner.

If choosing a gas tankless heater, be aware of venting requirements. Tankless gas heaters use a very hot flame to heat water, consume a coordinating amount of oxygen, and vent more exhaust. Read the manufacturer's venting instructions carefully and follow them exactingly. Electric tankless heaters, which require rerouting the wiring to support them, do not need to be vented, but the electrical reconnections might need to be inspected by a licensed electrician before approval by your city or county building inspection unit. Alternatively, have the wiring rerouted by a licensed electrician experienced in tankless installation to cover your liability if your house should burn.

Because tankless water heaters are still relatively new, some building inspectors might not be "up to speed" on the venting and connecting requirements. Keep all manufacturers' specifications handy when it comes time for the inspection process, and be prepared for objections and confusion. Tankless water heater installation will be a learning experience for all involved.

Both Bosch and AquaStar make highly rated gas and electric tankless heaters, for large or small applications. For more information, visit: **www.builderswebsource.com/techbriefs/tankless.htm#selecting**.

Alternatively, you can investigate a solar water heating system. Use the next Case Study, provided by Charles Nadel of Advanced Energy Systems, to learn more about this energy-efficient method of producing hot water for your home.

CASE STUDY: SOLAR HOT WATER HEATING

Charles Nadel

advancedenergysystemsusa.com

Advanced Energy Systems

A client who lives in a coastal community contacted us with an interest in a solar hot water system. The residence houses two adults and one child. The house has an almost east-facing roof with a 32-degree pitch. The roof has shading issues and is in need of new shingles.

Three adults have an estimated daily consumption of 60 gallons of hot water (20 gallons/adult/day, ASHRAE recommendation). Two adults and a child would be expected to consume 55 gallons of hot water per day. Hot water requirements of this magnitude can be met by two racks of evacuated tube collectors (44 tubes) or two small flat plate collectors (3.5-by-7.5 feet). The evacuated tubes provide more efficient wintertime (low light, short days) operation while the flat plate collectors will provide more efficient year-round operation. The client decided to use the evacuated tube collectors. To avoid placing the collectors on the roof, which presented numerous problems, the client opted for a ground-mounted system. The ground-mounted system was designed to face due south and have an inclination of 42 degrees, providing for optimal year-round operation. The collectors were paired with an 85-gallon solar storage tank and a heat exchanger.

The ground mount was constructed of concrete-filled 8-inch SonoTubes with embedded deck anchors, upon which were placed 4-by-6-inch pressure-treated stringers. The evacuated tube collectors were installed on the manufacturer-provided flat roof installation racks per the manufacturers' directions.

CASE STUDY: SOLAR HOT WATER HEATING

Figure 1. (Top left) Installing the frame. Figure 2. (Top right) Installed ground-mount frame.

Figure 3. (Bottom left) Installed evacuated tube collector system

Because the installation is in a northern climate prone to freezing conditions, a food-grade glycol solution is used as the heat transfer fluid. All piping between the heat exchanger and the collector is half-inch copper tube encased in 1-inch thick neoprene insulation. The piping is routed between the collectors and the heat exchanger in a trench, and insulated pipe in the trench is encased in corrugated flexible drain pipe for protection.

CASE STUDY: RURAL RENEWABLE ENERGY ALLIANCE

Case Study Courtesy the Rural Renewable Energy Alliance, Minnesota

And Joel Haskard, Co-Coordinator, Clean Energy Resource Teams (CERTs)

Regional Sustainable Development Partnerships

Hot Water Heaters in Central Minnesota

The sun shines in every corner of the earth, making it one of the most readily available sources for renewable energy. The Rural Renewable Energy Alliance (RREAL), a nonprofit committed to making renewable energy technology available to people of all income levels, is working to capitalize on the power of the sun to help four low-income families in central Minnesota.

Founded in November 2000, RREAL started a program that installs solar powered hot water heaters in low-income households at little or no cost to the family. Jason Edens, founder and co-director of RREAL, sees dual benefit to this work.

CASE STUDY: RURAL RENEWABLE ENERGY ALLIANCE

The solar hot water heat project in Central Minnesota began in September 2001 and ended in October 2003. During that time RREAL worked with Rural Minnesota Concentrated Employment Program Wadena-Ottertail Bi-CAP, Cass County Courthouse, Pine River Backus School District, Cass County Materials Exchange Program, Youth as Resources, Pine River Chamber of Commerce, Innovative Power Systems, and the University of Minnesota to ensure the success of their solar hot water heat project.

The objectives of the program are to reduce fossil fuel use, create greater energy self-reliance among four low-income families, create greater awareness solar thermal technologies, and create meaningful involvement of local "at-risk" youth in the solar system design and installation.

First, most families in Central Minnesota have electric water heaters. Electric hot water heating can consume up to 20 percent of a family's monthly energy budget, so solar hot water heaters save money. Edens estimate that a properly installed and sited solar hot water system will provide for about 70% of a household's hot water needs each year. Second, using a solar hot water heater decreases the consumption of electricity produced by nonrenewable fuels, having a positive impact on the environment.

RREAL has received funding from the Central Regional Sustainable Development Partnership to finance the projects, which cost between $1000 and $2500 per household for materials. The labor to install the systems is provided by volunteers.

The organization is able to keep the materials cost low by working closely with the Materials Exchange Program – a state program connecting businesses and organizations that have reusable good to those who can use them – to recycle used solar thermal equipment.

In a home without a solar hot water system, cold water is fed into the water heater to be warmed. When a solar hot water system is installed, it is sited between the water line feeding the hot water heater and the heater itself. All water that goes into the hot water heater is preheated with the energy from the sun, reducing the need for the electric or gas hot water heater to heat it.

The system itself works like this. Solar panels on the roof collect energy from the sun and use it to heat fluid, in this case non-toxic food grade antifreeze called propylene glycol. The heated fluid then runs through the house and is used to heat water through the use of a heat exchanger often located in the water storage tank. If the system was installed as part of the solar assistance program, RREAL provides upkeep and maintenance services to the family free of charge; though maintenance and upkeep are minimal with these systems. RREAL also installs and maintains systems for families

CASE STUDY: RURAL RENEWABLE ENERGY ALLIANCE

and individuals with the ability to pay for them. This helps RREAL finance some of its nonprofit activities.

Another exciting aspect to the work that RREAL is doing is their incorporation of area youth in the installation process.

Working closely with the Cass County Corrections and Pine River-Backus School District, RREAL uses the skills and energy of area at-risk youth in the installation of these units, while giving them real knowledge in renewable energy installation and maintenance. It is a win-win situation for both parties.

In addition to the benefits cited earlier, the project serves another very important purpose. It is educating area citizens and piquing their interest in renewable energy technologies. While most of RREAL's work has been primarily in Central Minnesota, they have collaborated on projects outside of the area. Ultimately, Edens hopes that the work of RREAL will encourage more citizens to use solar power and other renewable technologies in their daily lives. He hopes for the next round of installation they have will be able to host Community Education classes at the installation sites. In this manner, they can raise capital for future projects, as well as raise awareness about solar thermal technologies and installation techniques.

For more information, contact Jason Edens at 218-587-4753\

Net Metering

When you choose to install alternative energies, such as solar or wind, you might find days when your electrical production exceeds your need. If you live off the grid, you can store the excess electricity in batteries, for a limited amount of time. If you are grid-tied, you can sell this excess electricity back to your utility company.

In July 1983, the United States Supreme Court upheld a Federal Energy Regulatory Commission standard that allowed utility companies to pay no more than "avoided costs" for cogeneration, or between one and $.02 per kilowatt-hour. Since kilowatt-hours cost about $.10, most people did not connect their systems to the grid, and excess electricity not stored was simply lost.

The Energy Policy Act of 2005 (**www.doi.gov/iepa/EnergyPolicyActof2005.pdf**) mandated all public electric utilities to provide net metering, on request, to their customers. Newer, state-mandated net metering laws pay the consumer the fair value of electricity produced. Additionally, net metering allows the consumer to take advantage of lower electricity costs. For example, a local utility may bill you $.08 per hour for the first 100 kilowatt-hours, and $.10 thereafter. If you never use more than 100 kilowatt-hours in a month, you will always be able to take advantage of the lower rate. These rates are established on a state-by-state basis by the local Public Utility Commission.

To operate a net metering system, your utility company might have to change out your existing meter. Newer, digital meters do not run backward to account for electricity fed into the grid, but older, clock-style meters do. If you are grid-tied and returning power to the grid, your old-style meter will run backward when you are producing excess electricity.

Most utilities will "bank" your excess production and apply it as a credit to your bill, returning it to you at lower cost when you are not generating electricity. There is, so far, no blanket federal law that requires utilities to refund you in actual dollars. Some utility companies, however, will do so, at the end of a calendar year, for example. When you request net metering, check with your local utility company to see how it handles refunds under net metered production. Also, check with your local Public Utilities Commission to see if other companies subscribe to net metering and what their rates and policies are.

If you connect your alternative energy system to an existing grid, you will have to enter into an interconnection agreement as well as a purchase and sale agreement. Some utility companies that are into green energy themselves have developed simplified interconnection agreements. Others will require you to run the full gamut of forms, including federal standards forms and permissions.

Net metering is a relatively new development, spurred by an increase in affordable renewable energy systems, rising costs for electricity, and Renewable Portfolio Standards instituted by both the EPA and the Department of Energy. These standards will soon mandate as much as 20 percent of a state's energy supply come from renewable sources.

Net energy meters run forward when you are buying power from the utility and backward when you are producing excess power. Net energy meters also help the utility and federal agencies determine how much "green" power is being sold, as opposed to how much "brown" — coal or gas-fired — power is being bought. The results of net-metering incentives give the EPA and the Department of Energy accurate figures on which to base future renewable energy mandates.

If you are grid-tied and lack battery backup, you might not have power after a storm because your system is dependent on the grid, and the meter, for delivery. If you want to tie to the grid for the financial incentives, you should also consider batteries. Also, do not confuse programs such as Saver's Switch, or other interruptible power supply regimens utility companies use during peak loads to prevent overload, with green programs. These interruptible systems, called DSM systems, for Demand-Side-Management, benefit the utility company by preventing brownouts and blackouts due to excessive demand and may offer limited financial incentives but do not support renewable energy. In fact, they promote and endorse obsolete transmission and distribution systems by allowing utilities to ignore the problem of aging infrastructure simply by reducing peak loads on the system, loads which would otherwise force them to replace aging lines and poles.

Renewable Energy Credits

You, the consumer, can now buy renewable energy credits (RECs) even if your local utility does not have renewable energy as part of its production package. But what are RECs, and why should you invest in them? Because the extra few cents you pay for a kilowatt-hour of electricity goes to support research into and development of renewable sources of energy such as solar, wind, hydro, and geothermal. Without development of these resources, peak oil, which is just around the corner if not already here, will leave this nation dark, cold, and hungry.

There are two aspects of renewable energy funding. The first is green pricing, in which you, the consumer, agree to purchase "green" credits from your utility company, which in turn agrees to either sell you an equivalent amount of "green" energy, from wind, solar, or hydro generation. Xcel Energy of Minnesota, for example, offers green credits for wind from its Lake Benson wind farm. In Washington, you can purchase hydroelectric power. Because the costs are higher, the extra money you spend goes to promote and extend renewable energy. As of 2006, there were more than 600 participating utilities in 36 states offering green pricing options. To discover if your utility participates, go to: **www.eere.energy.gov/greenpower/markets/pricing.shtml?page=0**.

A participating utility company will charge you a higher price per kilowatt-hour, charge you a fixed monthly fee, or even offer residential and small commercial renewable energy systems for lease or purchase.

Green marketing, which is more competition-oriented, gives you the option of choosing between sources such as wind, solar, hydro, or geothermal. If you think hydro has the most promising future, you can find a green marketer who focuses on hydro. Green marketing so far lacks the impetus of green pricing because of transmission interconnection rules. The American Wind Energy Association (AWEA) is trying to ameliorate this situation by improving access to transmission interconnection and delivery for wind and other forms of renewable energy. When open access to transmission and distribution becomes the rule rather than the exception and becomes free of tariffs, green marketing will expand rapidly. Power companies who have so far focused on the bottom line at the expense of both the environment and renewable energy oppose open access, and these power companies have lobbying clout. But renewable energy, far from being a luxury, will soon become a necessity. For more information on this fascinating but complex subject, visit the AWEA's Web site, at **www.awea.org/policy/greenprins.html**.

Many people are unsure of the difference between green pricing and green marketing. One form of green marketing is purely corporate. For example, GE, a leader in the field of green marketing, offers light bulbs just like Sylvania, but its green profile is aimed at capturing a greater market share of the light bulb business. This is one aspect of green marketing at its best, because it makes corporations environmentally responsible and rewards them in rising sales figures for considering the environment. A corporation that pretends to be environmentally responsible but behaves in direct contravention of that principle is said to be engaged in "greenwashing."

Recycled Options

If you are building new or only remodeling, consider using recycled products to refurbish your home. You can buy new countertops made from pulverized porcelain sinks and toilets or made from glass recycled from automobile windshields, glass bottles, and old traffic lights. You can buy floor tiles made from pulp wood and cement, from recycled granite and river rock, or even from recycled porcelain and ceramic tiles. You can buy carpet made from recycled fibers, or lawn furniture made from recycled plastic, and both will help reduce future landfill requirements, though their manufacture may introduce some unwanted volatile organic compounds to your home's air. Before buying anything for your new or remodeled home, check the tag and make sure at least a portion of the product is made from recycled ingredients. You probably will not be saving any money, but you will be saving the planet.

Products from Recycled Materials

C & A Floorcoverings, out of California, makes nylon carpet tiles from 43 percent recycled product, with a backing that is 100 percent recycled plastic. These carpet tiles use no wet adhesive during installation to protect indoor-air quality and last up to three times longer than conventional carpet. Cameo Fibers, from North Carolina, makes a 100 percent recycled carpet pad, available in densities from 24 to 40 ounces. CrocoTile, from DuPont, is a carpet tile made from recycled materials from DuPont's Carpet Reclamation Project. Black and grey colors are 100 percent recycled material, with other colors containing at least 25 percent of a post-consumer recycled product. No adhesive is required, and the tiles last longer than conventional carpet.

Homasote Company makes a 100 percent recycled underlayment, similar to particleboard, made from recycled paper and containing no formaldehyde or asbestos, that can be used to replace or upgrade your subfloor before laying tile or carpet. Forbo Linoleum makes a 50 percent recycled linoleum flooring.

Vitrastone of Colorado makes a stone sink from recycled local products. Green Sacramento, out of California, makes and sells countertop materials, ceramic tiles, and even drawer pulls and cotton insulation, all made from 55 to 100 percent recycled content. Its aluminum sinks and glass tiles are 100 percent recycled content. Florestone, also from California, makes sinks and shower stalls from 10 percent recycled plastic and other post-consumer products.

Reusable Lumber Company, from California, makes S-shaped clay roofing tiles from recycled product. Greencor Composites makes a door core from rice straw and formaldehyde-free resins, and the finished product closely resembles real wood. Therma-Tru Corporation makes doors from a proprietary formulation of recycled wood and plastic composite whose ingredients are 100 percent postindustrial/consumer product. Even Pella, one of the nation's largest window manufacturers, makes its sash and frame-cladding for windows and doors from 93 percent recycled aluminum and uses glass that is 15 percent postindustrial waste.

U.S. Gypsum makes drywall corners made from recycled product or entire sheets of gypsum wallboard that are 75 percent recycled content. BPB, a British company with a division in Canada, makes ProRoc and GlasRoc wallboard from 100 percent recycled content. Philips Lighting Co., an internationally recognized resource for light bulbs and lighting accessories, makes compact fluorescent lights using 100 percent recycled mercury. One hundred percent of the inner packaging is post-consumer paper product, and 35 percent of the glass in the bulbs is also recycled.

You can find indoor furniture made from recycled plastics and other products or made from bamboo, which is a fully renewable resource. You can find end tables and table bases made from recycled wine barrels. Barnoire, of Vermont, is a two-man firm that transforms the remnants of broken-down barns into custom-built furniture such as cabinets, armoires, and cupboards. You can also find new homes built from this old wood. Similar organizations around the country reclaim old barn and mill wood for reuse as furniture, flooring, kitchen cupboards, beams, walls, and doors.

You can buy textiles made from 100 percent environmentally friendly, woven jute or coconut fiber. Draper Knitting makes high-pile, jersey, napped fleece, and terry cloth fabrics from 25 percent post-consumer product for use in clothing, industry, and as covers for paint rollers. Foss Manufacturing, New Hampshire, makes non-woven textiles and fabrics from 10 percent reclaimed polyester derived from x-ray film.

Crown Corporation, located in Great Britain with a division in Colorado, makes recyclable Anaglypta SupaDurable and Anaglypta Armadillo wallpapers. These high-relief, embossed wall coverings and borders can be painted or finished to look like oak paneling, wood, plaster, or metal. Both models are made from 90 percent postindustrial cotton and can be recycled again.

As you can see, there is almost no building or remodeling product that cannot be found as a recycled product. From walls to wall coverings, sinks to plumbing pipe, floor underlayment to floors, even lampshades made from 100 percent recycled cotton and coffee cups made from recycled plastic, every product you buy can be made from recycled materials. Check the label before you buy, and always buy recycled.

If you have a contractor, either building your home or remodeling your kitchen, check to see if it is using recycled building products. If it is not, request it does so, or consider switching contractors.

Salvaged Products

In the '60s, when I built my first home, salvage yards were almost unheard of. Today, they flourish. In the past ten years alone, the number of salvage companies has tripled. What was once difficult to find has become commonplace, and in many locations, the cost of salvaged materials, even unique items such as claw foot tubs and hammered tin ceiling panels, is surprisingly reasonable.

I live in a large Midwestern city. Within 25 miles of my location, I can find at least 25 companies who use recycled products, recycle, or create products from recycling. You can find similar resources for your state by going to **www.buildingreuse.org/directory**. This site, sponsored by the Building Materials Reuse Association (BMRA) will direct you to salvage yards, construction companies who use or engage in salvage, or architectural salvage yards that recycle wonderful old items such as fireplace mantels or built-in buffets from century-old homes. Here, you can also find antique chandeliers, filigreed door knobs, stained glass, or metal steam radiators — an architectural focal point even when not working. You might also find Victorian moldings, porch columns, pedestal sinks, and even solid oak doors, all at a fraction of the cost of buying new. Most architectural salvage yards have a wide range of lighting fixtures, hardware, used tile, kitchen cabinets, curiosities from another age, and a wealth of garden art or other oddities.

Salvage yards first took off when states began mandating higher dumping fees at landfills in the '90s. Contractors and builders who engaged in construction and had leftover material, or contractors who engaged strictly in demolition found dumping fees seriously affecting their profit margins.

Enter the age of salvage. There are benefits to salvaging, including longer-lasting landfills, less pollution, and less use of resources, but there is a downside. Homeowners who want to use salvaged wood in their construction will find costs higher than should be expected for used materials; this is because salvaged wood used for structural purposes has to have a stamp of approval, and this can cost as much as $200 per inspector visit. Salvaged wood also comes in different dimensions than "modern" wood, which means it has to be re-milled. Mills are not always willing to undertake such a project because the waste is exorbitant, affecting their bottom line, and a single bent nail can ruin a valuable planing knife.

At this time, "virgin," or freshly milled, construction-grade lumber is cheaper than re-milled salvage, simply because of the subsidies behind the lumber industry and the complexity and cost of re-milling salvage lumber. However, as more salvage mills, like the Materials for the Future Foundation mill in Oakland, California, open, the cost of salvaged wood will become competitive. State or federal subsidies supporting salvage operations would balance out the unfair advantage lumber companies currently have, but government requires the impetus of consumer demand to act.

Where salvaged lumber is not used for framing, the costs are competitive. Wood decking, interior floor boards, and fencing materials do not require an inspection certificate, and salvaged wood can transform an ordinary wood floor into a timeless treasure, filled with the character of age. In fact, some homeowner/remodelers select salvaged wood for precisely that reason, because it matches the vintage of their homes.

Salvage is not just about looks, though. Whether it is a classic Victorian glass doorknob or a renovated claw foot tub, salvage saves the earth's resources. Salvage a porcelain sink, says one source, and you save enough energy to run a 100-watt bulb for 100 days.

Architectural salvage yards seem to fall into two categories; junk yards with broken windows and rust-stained sinks, and museum-type displays of vintage building and refurbishing materials. If you are looking to save money and have the patience, visit the first; you might come across a hidden treasure. If you know that you want a vintage oak church pew, in excellent condition, visit the second. In some instances, salvage yards will post a notice for demolition, and you can register to help demolish an older home, salvaging the items you need as you work.

If you own an older home, you can contact a salvage yard for demolition, and in some cases, the demolition company will discount its fee for salvage rights. You can even on occasion donate salvage for a tax write-off. To find out how, contact the Used Building Materials Association, at (877) 221-8262, or the Reuse Development Organization, at (317) 631-5396. To find a salvage store near you, contact the Habitat for Humanity ReStores, at (800) HABITAT (422-4828), or visit: **www.habitat.org/env/restores.aspx**. These Habitat ReStores are retail outlets selling quality used and surplus building materials at a fraction of the usual price, and the proceeds from these sales go directly to local agencies to fund Habitat homes within the community.

Freecycle Network, a grassroots, global organization of 4 million members, are giving, and getting, free stuff for themselves and their communities. Begun in 2003 by Deron Beal, the organization is devoted to sustainability, as people from all walks of life band together to turn "one man's trash into another man's treasure." Visit them at: **www.freecycle.org**.

Older homes are notoriously difficult to remodel, because so much retrofitting is necessary, but you can do an energy-efficient remodel if you use salvaged products and stick to some basics. Next is a Case Study of a 1950s California home that was upgraded to exceed current efficiency standards.

CASE STUDY: KAREN FEENEY

Karen Feeney

Allen Associates

Santa Barbara, CA

1950s tract home remodel — energy efficient, healthy, and affordable

Project Background

"This home is an excellent example of what can be done to improve the comfort, energy efficiency, and indoor air quality of an aging San Roque tract home. Built in the early 1950s, this home was originally constructed without any insulation, making it extremely warm in the summer and cold in the winter.

"Construction techniques at the time did not reduce the potential for moisture problems down the road, which, along with the notorious San Roque clay soils, resulted in a number of water intrusion issues that needed to be addressed.

"The owners, who have a strong respect and appreciation for the environment and see themselves as 'early adopters' of green building technologies, were very motivated to explore creative solutions to these problems. Conserving energy was an important goal. A desire to protect the health of their two small children also played a significant role in how they approached their remodeling project."

Karen Feeney, Green Resources Manager, Allen Associates

Design Analysis

In this project, we used our Thermal Simulations to assess whether our current design could meet our energy efficiency expectations. We verified these simulations

CASE STUDY: KAREN FEENEY

by using sensors to monitor the house temperature and humidity throughout the year. This project also used many green products, achieved outstanding waste diversion results, employed LED lighting, used 100 percent of an on-site Acacia tree, and employed many other interesting techniques and solutions. We will organize them by category.

Planning

You cannot plan for everything, but be as prepared as time allows.

The Planning for PDR started in December 2005. Construction did not start until December 2006, and the project finished in September 2007. We spent one year planning, and I still felt like I did not get enough time to properly plan. I could have used another six to twelve months. Once demolition/construction starts, everything happens "quick," if only because there are so many decisions to be made. Spend the time in advance to think, plan, and design the elements that are most important to you. Discuss these early with your architect and contractor.

Below are some of the big-picture issues that we wanted to think through:

- Challenges to sustainable construction
- "Model" green options
- Construction/demolition goals

Sometimes what we see as a "problem" really is a solution. For example, the aging Acacia tree was nearing the end of its "safe" life; they tend to fall over as they get older and weaker. We decided to make the best use of what we now saw as an asset. We used the leaves and small branches as mulch for the very clay-like soils and milled the trunk and largest branches into the wood for all our cabinets, bookshelves, and bar top.

The beautiful back patio tile job could not be integrated into the new design, so we had it cut into 2-foot-by-1-foot blocks and made a bench seat in the back yard and used much of the demolished concrete to help landscape the back yard.

Site Orientation

Site orientation is critical to efficient design. There are many reasons to consider "site orientation," but the one I have focused on is a site orientation that contributes to a passive solar design. In other words, an orientation that allows for the sun's energy to positively influence the house system.

You do not have to have a south-facing property. South-facing property +/-15 degrees

CASE STUDY: KAREN FEENEY

is the ideal orientation. But, there are other orientations that can be advantageous to the design of a passive solar house system.

At PDR, the original house line faced largely east-west. This was ideal for our goal of trying to harness the sun's energy to warm and illuminate our house as well as cool the house on summer evenings by taking advantage of the prevailing winds. We also wanted to use solar thermal (hot water) and solar electricity (photovoltaics).

Sometimes you will have little opportunity to maximize site orientation, but if you are starting with a lot, looking for a house, or remodeling, you have a good opportunity to make nature work for you, or at least to create a partnership.

Mature trees and other aspects of the landscape are also important in creating an energy-efficient design. Shading the west side of the house from afternoon sun can help keep the room temperatures moderated.

Simulations & Analysis

Experiment with Different Design Solutions & Costs Before you Build

For our house remodel, we really wanted to have a better idea of the results or effects of our energy efficient improvements. We needed a tool that could a) help us learn by trial and error and b) give us data-based feedback on specific design improvements.

This is why we developed our own thermal simulations, which provide both of these capabilities.

A House "System"

There are a number of parameters or variables that can be investigated to simulate how a house will perform thermally:

- Window R-values and solar heat gain coefficient (SHGC)
- Insulation R-values, infiltration, air changes/hour (ACPH)
- Site orientation, number of degree days
- Amount and location of glazing
- Thermal mass: type and amount inside house

Because we were remodeling, we had certain constraints on our design. For example, our north wall was going to stay 2-by-4 framing while the rest of the house would be 2-by-6 framing. We opted for the larger framing so we could obtain a higher R-value in the walls with more insulation. We did have great southern exposure and wanted to

CASE STUDY: KAREN FEENEY

maximize that by capturing and holding the incoming energy of the sun. How would we do that with a raised foundation (no cement slab to act as a thermal mass)? Well, we intended to use a new kind of wallboard, phase change wallboard, that is able to hold and release energy over the course of a day. We will come back to that issue.

In the end, our thermal simulations demonstrated that for our location, climate, and site orientation, the windows and insulation would be critical in providing an efficient building envelope. In fact, a well-insulated building with high-performance windows was our minimum design criteria. Our analysis accounted for the size, location, R-values, and SHGC of all the windows; the types of insulation used in the walls, floors, and ceilings; solar insulation at the site; and air changes/hour, as well as the different types of thermal mass in the house. Our craftsman-style house had plenty of wood in the design, including the very dense Acacia wood, in addition to the normal furniture, cabinets, tables, granite, tile, and so on. Our simulations showed us that this incident thermal mass would indeed help to moderate temperature swings in the house.

Design Analysis

We can look at a few of our simulations for PDR, called "runs," to help demonstrate things. See information below on the 1) Run Comparison and 2) Detail Sheet. It is probably easier to talk about these or answer your specific questions — send an e-mail to info@DoeringDesignandEngineering.com with your questions.

A summary sheet, or "Run Comparison," is prepared that graphically represents the relative change in performance of the different design decisions. A number of performance metrics are used.

A "Detail Sheet," with all the specific run data, is created for each design. The Details Sheet records all the relevant design decisions; graphs the 24-hour temperature and heat flow (Q) data; and tracks the changes in a critical set of metrics.

Energy Efficiency

There are many things you can do to improve energy efficiency.

You do not have to buy solar panels to be "green" or energy efficient. In some cases the capital investment in solar panels might not the best way to spend your money. However, solar hot water heating is one of the best green investments you can make. But, there are many things that any homeowner can do. This is a big subject, so we will take it one step at a time.

Windows

High-performance windows, with an R-value around 3.0, will significantly improve

CASE STUDY: KAREN FEENEY

your home's energy efficiency. You also want to understand the influence of the Solar Heat Gain Coefficient (SHGC) on the indoor environment. If you are trying to maximize a passive solar design, you might want to allow more energy into the house with a higher SHGC, say around .60. However, you do not want to overheat the house either. Thermal mass in the house will help to absorb this heat during the day and release it as the temperatures cool.

Insulation

Good insulation, like windows, is a must for an energy-efficient building envelope. In California, our Title 24 standard insures that an adequate amount of insulation will be used. Using more than the standard is worth doing if the additional cost is marginal for an improvement or if a reasonable cost is possible for a significant improvement. Our thermal simulations were able to demonstrate where the best return on investment was. There are many type of insulation to choose from to get a healthy indoor environment, a sound/moisture barrier, and a use of sustainable products.

We used cellulose (recycled newspaper with a fire retardant applied), Icynene (blown in foam), and some fiberglass batts. Ever seen a fiberglass batt installation? That is one of the reasons we wanted Icynene in our walls; it has a good R-value/inch and really limits infiltration losses. It is a water delivery system (so no greenhouse gases are released at installation), and now there is a soy-based formula.

Water Heating

I strongly recommend heating your water from the sun. It is one of the best returns on your energy-efficiency dollars. These simple, passive tanks on your roof will provide you with free hot water very quickly, due to the short payback period. I suggest coupling the solar thermal water heating with an on-demand backup water heater (tankless).

If you have a larger house and can find a knowledgeable installer, you might want to investigate using a boiler to combine your water heating into one system for domestic hot water (DHW) and space heating.

Space Heating

If we did things just right with our passive solar design, we would not need to think about space heating because our indoor air temperatures would be between 65° and 73°F year-round. Although this is feasible in some climates, most codes require some type of backup heating.

There are many options and issues to think about. For now, I will just list some options to get you thinking:

CASE STUDY: KAREN FEENEY

- Boiler for both DHW and radiant floor heating.
- Heat pump, either ASHP or GSHP (air or ground).
- +93 percent two-stage Energy Star furnace, smallest capacity.

Natural and Artificial Lighting

Lighting can make up a surprisingly large amount of a home's energy usage. It does not need to. Use daylighting techniques, fluorescent & LED lighting, and conservation; turn the lights off when they are not needed.

Skylights and windows on two walls in a room will help to make the most of natural light. A passive solar home positioned facing south or southeast will get a lot of summer afternoon sun from the northern windows. In harsher climates you would want to limit the home's exposure on the north side, but in a temperate climate in a traditional neighborhood, you are going to have doors and windows on the north side; use them to get natural light into the house so you can keep the artificial lights off.

Afternoon sun on the west side of a home can inject plenty of heat into a house, if there is plenty of glazing. Design with this in mind, and use landscaping as post-design strategy to moderate the effect.

Other

We might have to make some changes to how we do things if we want to be energy efficient. Besides the many things you can do in the house, there are things to be done outside.

- Walk, bicycle, or take mass transit when you can
- Demand better fuel economy in cars, trucks, and commercial fleets
- Think about how your daily routines could be more efficient

Insulation

Energy efficiency requires energy conservation.

Many different types of insulation could be used in a remodel or new construction. Most have similar R-values/inch, but there are many differentiating characteristics.

In PDR we used Icynene, cellulose, and formaldehyde-free fiberglass in the walls, ceiling, and floor (and a few interior walls) respectively. In addition, I sealed all gaps and penetrations with a 30-year caulk to limit heat loss from the interior of the house to the attic. It is also important to talk about insulation in terms of windows and doors,

CASE STUDY: KAREN FEENEY

since a substantial surface area of the walls will be covered by windows and doors. To create an effective building envelope, it is critical to have energy-efficient windows and doors as well as good weather-stripping around doors.

California's Title 24 Standard mandates a minimum energy efficiency for compliance. Shoot for the following R-values or better.

- R-30 in attics
- R-13 in 2-by-4 walls, R-19 in 2-by-6 walls
- R-19 under the floor

You need to decide how healthy and sustainable you want the material to be in comparison to its cost, as well as what the installation method is.

- Loose-fill
- Blown-in (wet)
- Batts
- Rigid panel

I did not want to risk a poor fiberglass batt installation in my walls. So, while fiberglass batts are the cheapest, and you can get them formaldehyde free, there are many alternatives. For example, there is also natural cotton fiber insulation made from recycled blue jeans, or straw bales.

Icynene is an open-cell polyisocyanate foam. It is water-blown for installation, 99 percent air and 1 percent material. It stops air infiltration while allowing water vapor to permeate, is non-nutritive to rodents and mold, will not sag or lose its R-value over time, and contains neither formaldehyde nor volatile organic compounds (VOCs). Convective air movement inside cavities is virtually eliminated.

Mechanical Systems

Many efficiencies are to be gained from mechanical systems.

How you heat your water is one of the biggest decisions for energy efficiency. Water heating is one of the biggest contributors to your energy usage and costs.

Solar Thermal

Although the solar thermal heater is not really a mechanical system (gravity delivery and pressure fed), I cannot emphasize it enough. We plumbed in an on-demand or "flash" gas water heater downstream of a solar thermal heater. It has a mixer valve

CASE STUDY: KAREN FEENEY

to heat up or cool down the water as necessary. It is very quiet and efficient, with computer-controlled ignition, burning, and safety.

Plumbing

We used cross-linked polyethylene tubing (PEX) instead of copper. The choice was difficult when weighing the environmental and health benefits, sustainability, and costs. Read about it at Wikipedia, and then consider the pros and cons of copper. The PEX tubing used a manifold system near every delivery point. My plumber used copper for the actual stub-out to sinks, tubs, and so on to make a more rigid connection. The PEX tubing makes for a very quiet plumbing system.

Space Heating Revisited

After designing a dual-plumbed boiler system, more DHW and space heating, and modifying that to a fan-coil based system, it did not make sense to pour much money into my heating system when my goal was to not have to use it. Plus, I was not comfortable with the complexity and costs of the design. Instead, we specified the smallest furnace possible. We used a Bryant two-stage, 93 percent efficient Energy Star system that is wonderfully quiet and needs to come on only in the low stage, which provides a gentle warming of the house. The building envelope does the rest to hold the heat in for a long time.

Whole House Fan

We added a whole house fan in the centrally located hallway. It has many rpms, dual fans, and an R-22 automatically operated lid. It was exciting turning it on the first time. It is called a Whisper something or other, and it is quiet when it gets up to speed, but I love to turn it on for its effect. With the windows cracked, you can quickly exhaust the hot air out of the house as you draw cool air in.

Reduce, Reuse, Recycle

It takes some time and a creative effort to be thoughtful about all the waste generated from a significant remodel.

Landfill Diversion

In Santa Barbara, we have the benefit of Marborg Industries, a leader in recycling waste. It has a substantial sorting facility right in town.

We had a goal of 95 percent of the remodel waste being diverted from the landfill, and while that was a bit ambitious, we did reach 85 percent.

We used an aging Acacia tree that was on site for mulch and to build the frames

CASE STUDY: KAREN FEENEY

for all the cabinets throughout the house. We cut it down with a local crew, milled it in San Luis Obispo, 45 minutes north, and the cabinets were made in Somis, about 45 minutes southeast. While this was a risk to take since no one had worked with Acacia for cabinets, it is one of the parts of our project that I appreciate and enjoy the most.

Landscaping

We saved mature plants and trees, transplanting them to other parts of the yard. I am most pleased with the successful transplant of two 15-foot Eastern Redbud trees; now we have a mini-grove of three Redbuds on the southeastern edge of the property.

The Western Redbud is a beautiful deciduous tree with large purple leaves that change colors and drop. The new leaves are preceded by tiny, white-pink, elegant flowers along the branches. The leaves will get to be the size of a very large hand.

One man's junk...

The Internet is a great asset to quickly and efficiently get your stuff available to the masses. We regularly used CraigsList to find people who were interested in all sorts of stuff, including the old oak flooring. We also salvaged oil-rubbed brass lights, faucets, and sinks from a remodel in Orange County.

Habitat for Humanity, the Salvation Army, neighbors, and strangers were all helpful in relieving us of stuff that otherwise would have gone to the landfill. The wood waste was used at other job sites. Even our dumpster was combed through by strangers looking for "scraps.

It takes some effort, preparation, and thoughtfulness, but it can be fun and save your landfill at the same time. We sent our old windows to neighbors and an antique dealer. The cabinets were reused in the garage; the tub, wall heater, and doors were given to Habitat for Humanity. The fridge and washer and dryer went to the Salvation Army. Bookshelves were reused in the office. Some old attic insulation was reused in the garage and in isolated new foundations. The old oak floor was sold on CraigsList.

We also reused products ourselves, as our sinks and faucets came from an Orange County remodel.

New Construction

Consider the traditional "stick-built" home in which the floors are made of 2-by-12 lumber, the walls of 2-by-6 boards, and the roof of trusses. With the correct amount of insulation, this type of home can meet energy efficiency standards, but that does not mean it is eco-friendly. It requires an inordinate amount of wood and other resources, and the cutting of trees is never ecologically advantageous. Aside from the decimation of a forest, wood-harvesting requires thousands of gallons of petroleum, as chain saws wreak their havoc, trucks haul the cut wood to the mills, and the mills process it using thousands of kilowatt-hours of electricity. Add to that the cost of shipping the lumber from a forest-rich location, such as Alaska or Minnesota, to a region like Arizona, and you begin to see the scope of the problem of traditional building methods. If not for the subsidies the lumber industry receives, you would likely be unable to afford the wood for a new home on the average salary.

There are other, more earth-friendly ways to build a home. Some might fall outside the traditional building-permitting parameters of your location, but all are proven building methods. You can work with your local permitting agency to expand the parameters, or you can choose to build in a location where the permitting process is more lenient. You are not confined to "sticks;" you can use stones.

For more information on green building, visit Architecture 2030 at **www.architecture2030.org/home.htm**l, Green Home Building, at **www.greenhomebuilding.com/index.htm**, or the United States Green Building Council, at **www.usgbc.org/Default.aspx**. The United States Department of Energy has an energy efficiency and renewable energy online site, as does the EPA.

Straw-bale Homes

In the 1800s, settlers on the prairie used materials indigenous to the area. Those in wooded areas built log cabins. Those who settled where wood was not readily available built "soddies," a dwelling dug into the earth and roofed with prairie sod. Settlers in Missouri and Nebraska used baled straw to erect their homes. This building technique has been resurrected by more modern pioneers, whose straw-bale houses are not only eco-friendly, but also highly energy efficient and attractive. An 18-inch bale has an R-value of 48, almost three times as much as a 2-by-6 wood wall filled with standard, loose-fill, fiberglass insulation, which has an R-value of 2.2 per inch of thickness. R-values are derived from U-values, which describe how well a building material conducts heat at 75 degrees Fahrenheit with 50 percent humidity and no wind.

One California study indicates a super-dense straw-bale wall could save as much as 75 percent of traditional heating and cooling costs. The corresponding reduction in greenhouse gas emissions, and your carbon footprint, is huge.

Straw-bale homes are also cheaper to build in terms of labor costs. Stacked like enormous bricks, straw bales are easily assembled into walls without much building experience and require few power tools for their assembly, another savings in terms of energy. This ease of construction means a small home can be erected in a short time.

Straw is a waste product, the stem left standing after harvesting wheat, oats, barley, or rye. It is sometimes left standing in the field but more often is burned to provide space and nutrients for next year's crop. Unlike hay, which can be fed to cattle in large quantities, straw has little or no food or caloric value and causes gut compaction. Straw's main uses are as mulch in horticulture, erosion control on steep banks by state highway departments, as bedding for livestock and small animals, and as food for small mammals such as guinea pigs and rabbits that require large amounts of roughage for digestive health. More recently, straw is being used in the manufacture of biofuels such as biobutanol. Straw can also be used in such diverse products as horse collars, hats, baskets, and the production of paper. It can also be used to make thatched roofs, which are becoming an increasingly popular "green" style of roofing. Even with all these uses, much straw is still simply burnt where it stands, representing an ecological loss.Loose straw burns readily. When tightly packed into a bale, it resists fire, as does a book with many pages. This happens because the straw in the bale is so densely packed it admits oxygen very slowly, and fire needs oxygen to burn. When combined with natural lime, gypsum, or clay plasters, this "fresh-air" effect is enhanced. Standard drywall, on the other hand, contains additives designed to reduce mildew and resist fire, and these additives contribute to the volatile organic compound load inside a typical home.

In addition to their thermal resistance, straw bales are also highly soundproof. If you lived in a straw-bale home next to an airport, the decibel value of airplanes taking off would be reduced by a factor of three.

There are two ways to build a straw-bale home. The first is a post-and-beam construction, in which wood, steel, or concrete is used as the framework and the bales are placed inside the walls to act as insulation. The second uses only straw bales, in which a top plate is laid and fastened to the foundation to support the roof. In either system, the bales are stacked like bricks and pinned with rebar, wood, or bamboo stakes. Chicken wire is wrapped, both inside and outside the walls, and attached firmly to the bales, forming a foundation for the plaster. A bale home can be plastered without the wire, but the wire also offers structural stability and is not very expensive in any case.

Loose straw burns readily. When tightly packed into a bale, it resists fire, as does a book with many pages. This happens because the straw in the bale is too densely compacted to permit the entrance of oxygen, which fire needs to burn. In fact, in fire safety tests conducted in New Mexico and Canada, plastered bale walls subjected to 1,850 degrees Fahrenheit did not ignite for two hours, or until a small crack formed in the plaster. Even then, the smoldering was easily extinguished.

Moisture is a problem in any home. In conventional homes, moisture can cause toxic buildups of mold and mildew. In the Pacific Northwest, some 60 bale homes were sampled for moisture in the walls. The average reading was between 8 and 10 percent. Moisture problems that have occurred may be the result of poor design and/or construction and include areas such as the foundation, roof, and plaster. With these problems resolved in advance, bale homes can last indefinitely. The Burke House, in Alliance, Nebraska, is a century old. Other historic bale homes built at the turn of the last century prove that bale homes, properly built and maintained, can have a useful lifetime of 100 years.

Rodents will not eat straw, as it has no food value. Insects may get inside the straw and nest if the plastering job is shoddy or incomplete, but this can as easily happen in traditional, stick-built homes. In a survey of the 600-some bale homes in existence, few owners reported any significant insect infestations.

The cost of a bale home is comparable to, or a little higher than, that of a traditional, stick-built home. The single factor affecting cost is the plaster. Homeowners who choose bale homes do not do so to save money but because they feel the increased energy efficiency, a healthier indoor environment, and the eco-friendly nature of straw-bale construction are their own rewards. A traditional house, with the insulation upgraded to match the R-value of a bale home, would cost the same.

In the '50s, it would have been difficult, perhaps impossible, to get financing and insurance for a straw-bale home. Times have changed. Bankers today are willing to lend, because they have seen the resale values of such homes and the quality and durability of their construction. Even if your local banker is unfamiliar with bale construction, you may already have built sufficient rapport to convince him your project is worthy of a loan. Insurance agents, once leery of such outlandish building practices, have also come around. They may not understand the more esoteric construction aspects, but they are more open-minded than they were. In some instances, they are simply more interested in the distance to the nearest fire hydrant. Both American Family, which currently insures more than 24 bale homes in Colorado, and Allstate are good places to start. Farmer's Insurance will reportedly insure a bale home at preferred rates. Even Fannie Mae and the United States Department of Housing and Urban Development, or HUD, are coming around to the concept of straw-bale homes.

Building inspectors and building codes, once far below the learning curve when it came to unusual building designs, have also adapted to the times. Codes have been developed in almost all states for both post-and-beam and load-bearing straw-bale construction.

There are two essential considerations when building a bale home. Never use synthetic stucco finishes, which can trap gases inside the bales, and always install a backup heating system — in addition to your solar application or wood-burning stove, as the Uniform Building Code standard requires it.

For more information on bale homes, visit: **www.epsea.org/straw.html**, or **www.balewatch.com**.

Adobe Homes

As a girl growing up in Southern Colorado, I lived in an adobe house. The house was cool during the hottest days, warm in winter, and smelled always of warmed earth. The light coming through the windows picked up the pale rose color of the walls, lending a soft glow to the rooms.

The Egyptians used adobe, and so did the South American Indians, who piled wet mud on dry, allowing each layer to cure before adding another. When the Spaniards arrived, they showed the Indians how to form the material into bricks. "Adobe" is a Spanish word with Arabic origins — specifically, "atob" — which means "sun-dried brick." Adobe bricks are earth, or soil, with a proportion of clay. The earth, mixed with water, is packed into a wood frame and allowed to set for several days, where it forms bricks. These bricks are set on end and allowed to dry for almost a month, under moisture-free conditions. Modern adobe bricks are 14 inches long, 10 inches wide and 4 inches high. They are still made with straw and may have a small amount of asphalt as well. An adobe brick costs about $.60 if bought from a manufacturer, or you can make them for nothing if you have access to a clay-type soil, water, a wheelbarrow, some 2-by-4s, nails, a shovel, strong arms, and time.

Adobe bricks are heavy and require a good foundation, dug in firm soil, below the frost level for your region. Foundations should be a minimum of 8 inches thick and 1 foot wider than the proposed wall, and they should stand above the ground about a foot. Because of the possibility of seismic failure, adobe walls should not stand higher than ten times their thickness. This means three courses of 4-inch wide brick laid side by side to achieve a standard 7- to 8-foot wall.

Bricks are laid with real mud, and it is best to lay them in warm weather to prevent the mud from freezing. Each course is laid the entire length of the walls simultaneously, with overlapping bricks making up the corners. Story poles, set at the ends of the walls and connected with string, are used to mark each additional course of brick. Window and door locations are topped by lintels as the courses are laid, serving as support for the openings. The whole lacks the precision of stick-built construction, yet adobe is sturdier than traditional construction and will resist high winds easily.

When the walls are high enough, beams are laid to tie all the walls together. After that, long poles, called "vigas," are laid to support the roof, with bricks laid between these along the tops of the walls. The vigas are covered with latias at 90-degree angles to each other, creating a lattice-work of smaller poles, usually made of pine or aspen. Traditional adobe roofs consisted of bricks laid over latia, topped by an inch or more of mud. Modern methods employ insulation, gravel and tar, and use 2-by-6 joists angled to provide a sloped surface for runoff. Both styles have canales, or drain channels, every 10 feet or so.

The ends of the vigas not only provide protection to the adobe walls and foundation, but also are a point of architectural interest.

Modern builders of adobe homes may also use foam insulation on the outsides of the walls and apply screen to help hold the exterior plaster, just as in bale construction. Plastering is a slow, precise process and requires three or more coats to achieve an effective finish. Here again, the primary cost factor in an adobe home is the cost of plastering. Plastering is both an art form and a building technique, lost to the modern world with the advent of cheap drywall. Even if you can find a journeyman plasterer, it will cost you dearly; these artisans make $35 an hour on the union scale.

Adobe bricks are inexpensive. If you build your adobe home yourself from reliable house plans, it will cost less than a traditional home. You can expect to pay $30 per square foot for a simple adobe, using salvaged materials and your own time, even if you incorporate those treasured nichos or bancos. Nichos are niches in the adobe in which to place statues or bouquets, and bancos are benches or shelves made of adobe which jut from the walls. If you have to hire labor, costs approach $65 a square foot.

If you contract to have an adobe house built, the labor costs will boost the price of your home to $100 per square foot, which is close to the cost of traditional building. A stick-built house can be completed in a matter of months; an adobe house could take three-quarters of a year. In a traditional house, about 5 percent of the space is taken up by walls. In an adobe home, the walls take between 15-20 percent of the total space because of their thickness. Adobe homes, however, retain heat and cold far better than traditional, stick-built homes, and adobe contains no volatile organic compounds to poison you or your family.

Adobe construction meets the Uniform Building Code standard and is easily insurable, being almost 100 percent fire resistant. It is durable, lasting hundreds of years in dry, arid climates such as the American Southwest and filled with a unique grace and charm that Southwesterners have come to recognize as their heritage. Farther north, where climates are wetter, rammed earth homes are more appropriate; more on them next. The nicest part about homes made from earth is that they are traditionally do-it-yourself projects, using materials that permit a few mistakes and some trial-and-error learning. Both are also a very "green" approach to building, contributing few if any pollutants to an already troubled earth and providing their owners with contaminant-free shelters that are easy to heat and cool, further reducing the owner's carbon footprint.

Rammed Earth Homes

Rammed earth building dates from the Great Wall of China. Settlers from France, Germany, and England brought the technique with them to America, where it was used both before and after the Revolutionary War, until mass-produced brick and lumber became widespread, readily available, and affordable. Thomas Jefferson built Monticello of rammed earth and vigorously promoted similar building techniques.

Rammed earth homes consist of walls made of suitable mixtures of sand, gravel, and clay. Pre-modern rammed earth homes used animal blood or lime as a stabilizer; the common stabilizer today is cement. Forms for the wall are set up, and about a foot of material is poured inside the form, then tamped down to about half its volume. The process is repeated until the entire form is filled. At this point, the wall is so solid the forms can be removed. The walls are allowed to dry for about 20 days and will continue to cure for up to 2 years. The longer the wall is allowed to dry, the stronger it becomes. Modern rammed-earth buildings may use heavy equipment to compact the mixture, and these walls may be 1 foot thick or more.

Exposed walls should be sealed to prevent water damage, with an environmentally friendly soy-based sealer such as Acri-Soy MST sealer. A similar, indoor sealer can be used on interior walls to prevent dust, a natural by-product of earth-based construction. Foundations should be adequate to keep the rammed-earth wall above snowmelt, or standing water left by heavy rainfalls, and eaves should be sufficiently deep to protect the tops of walls.

Like adobe, rammed-earth homes have a very solid feel and easily modulated temperatures, and they are very quiet. Rammed-earth walls are also vermin and fire resistant. They meet common building codes and are the most-seismic resistant of any building style currently in use.

Because of the structural stability of rammed earth, houses of two stories are possible without sacrificing structural stability. Rammed-earth techniques can also be used to lay a smooth, impermeable floor, with the addition of enough cement and extensive finishing techniques. Most rammed-earth floors are sealed with linseed oil, but you can also use one of the new, soy-based sealers. Because earth is not impermeable like stone, even when sealed, you may have to refinish your kitchen floors — or floors in other high-traffic areas — once every 5 years or so. Due to rapid thermal expansion, rammed earth may not be suitable for a fireplace, and stone would be a more appropriate choice. For instructions on building rammed earth homes, visit: **www.motherearthnews.com/Homesteading-and-Self-Reliance/1973-09-01/How-To-Build-a-Rammed-Earth-House.aspx**.

Okamoto Saijo Architects designed and built a rammed earth, passive solar house in Napa Valley, and the information is provided next.

CASE STUDY: RUBISSOW FARMHOUSE

Rubissow Farmhouse

Napa Valley, California

Okamoto Saijo Architecture

Project Description

Okamoto Saijo Architecture (OSA) designed a Napa Valley, California, residence based on passive solar design and inventive use of recycled building materials.

By focusing on these two concepts, the resulting design creates open spaces that seem larger than the net floor area and finish details that appear more expensive than a modest construction budget would normally allow.

The farmhouse is located at the end of a southwest-facing, forested ridge above the Napa Valley floor. The climate is not only ideal for wine grape growing, but also well suited for solar design — evening summer breezes help cool during the summer while solar radiation heats the interior spaces during the winter. A successful passive solar design includes a proper balance of thermal mass, south-facing windows, sun-shading roof eaves, natural cross ventilation, and a highly insulated building envelope. The major rooms are oriented south for optimum solar exposure, while the stained concrete slab floor and earth walls provide thermal mass. Cellulose insulation, which has higher R-values than ordinary fiberglass batts, fills the stud walls and ceilings. Windows afford a wealth of natural light to fill the spaces and frame the wonderful views of the Napa Valley outside.

The availability of new and innovative recycled building materials inspired creative

CASE STUDY: RUBISSOW FARMHOUSE

detailing and juxtapositions. The 18-inch thick PISÉ earth walls contrast with the horizontal fiber-reinforced cement siding on the exterior walls, while sheet metal flashing separates the siding from OSB wood panels. Inside, tall redwood doors made from an old water tank divide the private rooms from the central living room, while the bathroom vanity counter is made of recycled glass chips poured into a cement slab.

Project Data
Specific Service: Full Architectural Design Services
Project Architect: Paul C. Okamoto
Construction Cost: $170,000
Floor Area: 1,100 square feet (plus 100 square-foot loft)
Completion Date: Winter 2000
General Contractor: Geoff Austin, Leeward Construction

Insulated Concrete-formed Homes

Insulated concrete forms are 4-by-8-foot panels of rigid foam insulation made from expanded polystyrene, extruded polystyrene, or polyurethane. When erected vertically and connected, they create a form so strong and stable you can fill it with concrete. These forms are used as an alternate method to cement-block basements and — when the forms are left in place — provide highly insulated living spaces, either below or above ground. The strength of insulated concrete-formed, or ICF, construction is such that homes built in this manner remain standing even when exposed to winds greater than 200 miles per hour. Insulated concrete-formed homes are also reinforced with steel rods, known as rebar, making them almost twice as strong as free-standing concrete alone.

The insulating value of a 2-inch panel is R-12. When both inside and outside forms are left in place, the insulation value of insulated concrete-formed walls is R-24, or 50 percent greater than 6 inches of fiberglass. Because of this, insulated concrete-formed homes are highly thermally efficient, and the lack of air infiltration means mold and mildew will not become a problem. Insulated concrete-formed homes with basements also take advantage of the natural heating and cooling benefits of the earth — a constant 55°F at 4 feet below grade — and this effect provides additional temperature modification to all the rooms in an insulated concrete-formed house. In fact, insulated concrete-formed homes are so energy efficient, most manufacturers will guarantee a 50 percent savings on heating and cooling costs.

Insulated concrete-formed homes come in several types: Flat systems provide a uniform width of concrete throughout; grid systems produce a waffle-shaped wall, which is thicker at some points than at others; post-and-beam systems have individual horizontal and vertical columns of concrete completely enclosed in foam. All systems meet the Uniform Building Code standard. The forms themselves provide furring strips on which sheetrock and siding, or other interior and exterior finishes, can be affixed. Insulated concrete-formed homes have a three-to-four-hour fire rating and cost about the same as a stick-built house. Contractors estimate the life of an insulated concrete-formed home at 30 percent longer than traditional building. In an insulated concrete-formed home, you do not have to find a wall stud to hang a picture.

In spite of their energy efficiency, however, insulated concrete-formed homes do use a great deal of foam, and these foams are often made from polyvinyl chloride. This makes the home energy efficient but not as eco-friendly as adobe or rammed earth.

Steel-Framed Homes

From the outside, it is hard to tell a steel-framed home from a traditional, stick-built home. These homes use the same interior and exterior wall finishes as traditional building, and only their durability, the depth of their wall cavities, and the insulation therein distinguish them, but none of these features are visible in a completed home.

Unlike wood, steel does not settle, warp, rot, or come under attack by insects. Steel-framed walls built with heavy, steel members on wide frames are easily moved, meaning you can enlarge your home over the years without the prohibitive costs and mess associated with wood-frame remodeling. If you choose steel-for-stick framing methods, remodeling will be equivalent to remodeling a stick-built house.

Steel homes are very energy efficient, and much of the steel used to build them is recycled. In spite of this, the processes used for making or recycling steel are very labor and energy intensive, and steel-structure homes are not the most environmentally friendly.

Composite Lumber Homes

Composite lumber is made from recycled wood and plastic and pressure-formed with high-impact adhesives. It may consist of 50 percent sawdust or wood fiber and 50 percent waste plastic. The material is formed into solid profiles, such as decking lumber, joists, and trusses, or hollow profiles such as doors. Because the plastic encapsulates the wood fiber, composite wood resists moisture better than conventional lumber. Composite lumber is also heavier than conventional lumber and becomes more rigid or more flexible than conventional lumber, depending on the temperature.

Composite lumber reduces some of the impact logging and lumber production has on our forests. This, combined with advanced framing techniques, can reduce our dependence on wood.

However, due to existing building code regulations, the use of composite lumber may be restricted or prohibited, particularly in exterior, load-bearing walls where holes have to be drilled to permit access for electrical and plumbing. It still remains an excellent, environmental alternative when putting in a deck, porch, or gazebo, and trusses made from composite wood meet or exceed industry standards.

Post-and-Beam Construction

Post-and-beam construction dates back to the Greeks and includes any structure in which upright posts of any material support beams. Timber framing is a highly sophisticated form of post-and-beam construction, in which lengths of wood are cut to size, either on site or at the factory, and are notched, grooved, or otherwise carved to join two pieces of wood. The common joinery used is mortise and tenon. Post-and-beam construction, or timber-framed houses, use heavy posts and long expanses of beams, without employing the 16-inch, on-center studs common in traditional wood-framed homes, which means windows can be the width of the house, giving post-and-beam houses a lighter, more open feel.

Timber framing became obsolete in the middle of the 19th century with the advent of stick-framing, in which precut lengths of lumber were shipped to a site and assembled using nails or other fasteners. Only the Amish, and a few eccentrics, continued the tradition of timber framing until recently, when this building technique enjoyed a renaissance.

Post-and-beam or timber-framed homes are beautiful in a massive way, reminiscent of European cathedrals. The exposed beams and vaulted ceilings are magnificent, and the oversized timbers imply a strength not found in stick-built homes. One particular half-timbered house in France dates back to the 12th century. These houses can also achieve significant R-values using panels that wrap the frame, and their exteriors, composed of exposed beams and stucco, brick, or stone, have a classic appeal.

A timber-frame house will cost between $150,000 and $300,000, depending on your location and choice of wood. This cost does not include the foundation, the septic system and well or equivalent connections to local water and sewer utilities, or the cost of the land and the site preparation.

You can cut corners by keeping your home small and your floor plan simple.

A rectangular-shaped home is cheaper than an L-shaped home or one with wings. It costs more to build out than it does up, so incorporate a second story or loft into your design if you need extra space.

Timber-framed houses, no matter how elegant and stylish, are wood-framed and therefore not environmentally friendly like adobe or rammed earth. If you must build post-and-beam, at least choose a common, readily renewable wood, like white pine.

Log houses, which are fully timbered rather than timber-framed, commonly use saddle notches to join layers of logs and use a good deal of wood. The advantage of a log home is that little or no wood from the tree is wasted as it is in a planing mill, where as much as a third of the tree may end up as wood scrap, wood chips, or sawdust after milling lumber for stick-built homes. Log cabins blend well with rural landscapes and can be made thermally efficient with insulation. However, wood burns readily and is subject to infestation by termites and other pests. If you live in the country on ten acres of wooded property, you might be able to justify a log house, particularly if you build it yourself from your own trees. Otherwise, allow your trees to flourish, adding needed oxygen to the air and removing pollutants, and buy a log-home kit. To find a local contractor, visit **www.logassociation.org/directory/builders_us.php**. Always buy materials locally, as this saves on shipping and further reduces your carbon footprint.

Stone Construction

We have been building with rock ever since we left caves, and this choice is still the most environmentally friendly one available to us.

Laying stone to form a wall is both art and science, and it can be learned either by watching or doing, depending on your mind-set and preferred way of learning. One important item to remember is that overlapping stones make a stronger wall. Another is that the largest stones should form the base of the wall. Once you have these tenets fixed in your mind, you can pile up some stones, a bucket of mortar, and practice. Squared or flat stones are preferable, because they stack better, but round stones will work, too. Just use a stiffer mortar.

Rock has excellent thermal properties, retaining heat and cold equally well but should be insulated on the inside to keep the temperature in equilibrium. More important, rock is beautiful, implying a strength and permanence imparted by no other building material.

Rock walls can be laid by hand or slip-formed. Slip-forming involves a movable, wooden frame, commonly of 2-by-12s, into which stones are placed; the concrete or mortar is poured over the stones, and the form is moved after the wall segment has dried sufficiently to be stable. At this point, the excess mortar would be removed from the stones to reveal them. This can be done with a wire brush and small amounts of water.

Rock walls require immensely durable and stable foundations; the weight is enormous. Dig well below frost line, tamp the earth firmly, and make your foundation wider than the wall, as you would with an adobe home.

Rock withstands fire, is insect resistant, does not support the growth of mold, and is hypo-allergenic. Additionally, rock does not require the use of non-renewable and polluting resources such as petroleum to manufacture.

Earthship Homes

Michael Reynolds, the founder of the Earthship concept, lives near Taos, New Mexico, and has his own business called Earthship Biotecture. Reynolds has devoted his life and his Web site (**www.earthship.net**) to the furtherance of Earthship building techniques, which are becoming more popular as environmentally conscious consumers look for ways to reduce their carbon footprint and live in harmony with the earth.

You can buy a packaged Earthship home, in sizes ranging from 600 square feet to a three-bedroom, two-bath model, comprising 1,600 square feet. Earthship walls are commonly heavily earth-bermed structural walls with passive solar architecture; that is, the glass wall of the house faces south in northern latitudes to take advantage of the heating capability of the winter sun, and north in southern latitudes, to prevent the house overheating. In addition, Earthship homes feature a canted southern wall, angled so the surface is perpendicular to the angle of the sun in winter, permitting the wall to absorb heat during the winter. The thickness of the walls further prevents the loss of heated or cooled air. The structural walls may be constructed with used tires filled with rammed earth and stacked like bricks. The interior surface is plastered with adobe or cement. You can also select structural walls made of straw bale, concrete, or adobe. You can buy construction drawings meeting the Uniform Building Code standard for about $5,000, and home kits can be delivered anywhere and adapted to any climactic extreme. These houses are so thermally efficient that temperatures rarely exceed 65 degrees, even after two days of clouds and one night at 17 degrees below zero, without a backup heating system.

Earthship homes feature solar or wind-powered electricity, heating, and cooling, often with battery backup; on-demand water heating; rain-barrel, or cistern, water catchment; distributed grey-water and black-water treatment facilities, which recycle usable water; composting toilets; and even indoor gardening options. The only objection I could discover was the use of rigid insulation around the exterior to protect the wall mass and stabilize interior temperatures. However, if you can find a way to use some of the more environmentally friendly foams, such as Icynene, you will have superior energy efficiency without pollution. All the options mentioned above are also adaptable to conventional approaches, such as grid-connected electricity or gas hookups, utility-provided water and sewer facilities, wood stoves, and fireplaces.

Earth-rammed tires are the most economical way to build an Earthship and offer the most structural stability, but you can use almost any waste product; Earthship is a concept, not a building code. Reynolds has used aluminum cans filled with concrete, but you could potentially incorporate any solid waste product, such as bottles, glass, and rigid plastic from electronics. Just check your local building codes before beginning. Climate and building codes are the two most relevant aspects of Earthship design; failing to take them into consideration will result in a thermally inefficient, unsound structure.

There are Earthship homes in almost every state in the United States and in many European countries as well. Earthship homes offer high thermal mass and may be earth-sheltered as well, meaning earth is piled up outside the structural walls, or the structure is built partly or fully below ground level. The latter is the most thermally efficient style of building, as it allows the building to take advantage of stable temperatures as little as four feet below the surface.

Earthship homes capitalize on energy efficiency, sustainability, and ease of building. You can build one yourself, if time is not a consideration. All you need is a basic knowledge of plumbing, carpentry, and wiring.

There are drawbacks to an Earthship home, too. The first is their resale potential. You may have to search long and hard to find a buyer who also values sustainable building. The second is building permits. Rammed earth has its own code now, passed in New Mexico in 2004. If you live in another state, you might have problems convincing your local building inspector that rammed earth is stable. Financing is another hurdle, best crossed by you and a banker or loan officer with whom you have built rapport.

If you can jump through the 10,000 hoops presented by this new green building technique, you will have a home that used only 10 percent of the energy required to erect a stick-built home, and you will use only 10 percent of the energy required to sustain that kind of home.

Cast-Earth Construction

Cast Earth, one of the newest developments in earthen construction, is somewhat more affordable than Earthship, rammed earth, or adobe and cement constructions. Cast Earth walls are composed of earth, 10 to 15 percent calcined gypsum, which is gypsum baked to a powder, and other additives poured on site into a frame, like insulated-concrete homes. Calcined gypsum has a crystalline lattice structure that gives the resulting Cast Earth walls a great deal of strength. According to its inventor, Cast Earth consistently tests above 250 pounds per square inch, which is more than twice the strength of adobe.

By itself, calcined gypsum sets in about 30 minutes. Combined with earth, it sets in less than ten minutes. As a result, a Cast Earth house can be put up in 36 hours. Cast Earth is blended in concrete mixers and then poured into metal forms, but any form material is adequate, including rigid foam, enabling the builder to make curved or angled walls as required.

The stem wall is poured first, followed by two to three feet of wall section. When these sections have dried all around the perimeter, more Cast Earth is poured until it reaches the tops of the forms. The Cast Earth mixture can be colorized, by the addition of iron oxides for example, left plain, or painted with a colored sealer, giving Cast Earth the appearance of adobe.

Although gypsum provides better moisture resistance than adobe, Cast Earth walls still need to be sealed and provided with a good roof overhang to prevent water damage. Cast Earth has inherently lower labor costs than adobe and the structural solidity of poured concrete at a fraction of the cost. Cast Earth homes, properly sited and planned, cost half the amount of adobe.

Cast Earth does not shrink as it dries, like adobe does, so it is much less likely to develop cracks. Cast Earth can also be made from a wide variety of soils, whereas adobe is restricted to soils that contain clay and larger aggregates in precise proportions, and cement will tolerate fine, clay particulate in only minute proportions.

Cast Earth, invented by chemist Harris Lowenhaupt of Arizona, is still in the development stage awaiting patent approval, but expect this to be the next big building fad within a few years.

For more information on Cast Earth homes, visit **www.greenhomebuilding.com/cast_earth.htm**.

Papercrete Homes

Landfills in the United States are filled with almost 50-percent paper and paper-type products (food and beverage containers, cardboard, fiber planting pots, toilet paper, etc.).. Paper is supposed to biodegrade in a matter of months. Unfortunately, as landfill owners and inspectors have discovered, compacted layers of paper do not degrade for years. Newspaper pages from the '70s have been found in which the text is still completely readable. Computers, which were supposed to usher in an era of "paperless" documentation and correspondence, have failed to live up to their promise.

Papercrete, one of the newer ingredients in the building world, might solve that problem. Mixed with Portland cement in precise proportions and under exacting conditions, Papercrete forms a unique building material, sometimes called "padobe" or "fibrous cement," that represents sustainable building and a welcome relief to landfill capacity. Papercrete blocks look very much like cement blocks but are much lighter, have a tensile strength of 260 pounds per square inch, and — at a common thickness of 16 inches — provide an R-value of 32, or twice as much as a stick-built home that meets current insulation codes.

Papercrete is paper, preferably newsprint but also glossy paper and cardboard, that has been "re-pulped," or soaked in water and blended with Portland cement and/or clay or dirt. Some users have experimented with the addition of sand, powdered glass, rice hull ash, and even fly ash recaptured from power plant smokestacks.

Invented and patented in the 1920s and discarded as less than profitable, the concept of using paper to make building blocks was resurrected by Eric Patterson and Mike McCain, both of whom have contributed considerable time and effort to finding new ways of making Papercrete and using it to build.

Papercrete is very inexpensive, costing an average of $5 to $10 per square foot, as compared with $50 to $80 or more for a very simple, stick-built home. However, Papercrete is still an experimental building method, and mistakes have been made. In spite of that, Papercrete homes, properly built on good foundations and protected from the elements, show promise of lasting as long as adobe.

One of the most eco-friendly aspects of Papercrete is the way the paper fibers bond to Portland cement. When water is added, then drained from the mix as the block cures, the water, filtered through the paper, comes out almost completely free of cement residue. Papercrete is hydrophilic — that is, it holds water — but if properly blended and installed, it will also allow that water to drain away without losing its shape or integral mass. There are partial-Papercrete homes in such damp locations as Vancouver, Wisconsin, and in Massachusetts, where one house, built in 1924, still stands firm. Given this, humidity might not be as great a factor as once thought. However, given Papercrete's experimental status, homeowners who wish to use it should select dry areas rather than floodplains.

Papercrete is dimensionally stable in a wide range of temperature and humidity, is relatively fire-resistant, repels rodents and insects, and can be nailed or screwed. It will absorb moisture, particularly from the ground, and will become moldy if it remains damp for an extended time. Papercrete might even deteriorate if not protected.

Cob Homes

Cob is a very ancient building method involving claylike earth, straw, feces, and other fibrous materials mixed with water. In medieval England, the water and solids were laid on the ground, trampled into a usable mixture by oxen, then poured into rock-wall forms and packed down by workers' feet in a process called "cobbing."

Cob walls may be about 2 feet thick, providing excellent thermal qualities. Cobbed houses have been known to survive more than 500 years, and some are still habitable. The walls of cob homes are very free-form, giving the builder ample opportunity to provide a unique, architectural focal point. In damp climates, cob must be protected by deep roof overhangs, high foundations, and sealing or stucco/lime plaster, just like other earthen structures. Unfortunately, cob homes face the same drawbacks as other unconventional building practices, namely resale potential, financing, and permitting. At this time, no Uniform Building Code standard exists for cob, so check with your local permitting department.

Cordwood Homes

Stacked cordwood has excellent thermal properties, due to the lengths of cordwood, and significant mass as a result of the mortar used to bind the wood but much structural stability.

After you have laid an adequate foundation, keeping in mind that cordwood is not as heavy as earth, install posts topped by beams. Lay cordwood in the walls with the ends facing out, mortared together with cement, the ends projecting from the mortar about an inch. Pieces of wood can be from 12 to 24 inches long. The standard cut for cordwood today is 20 inches, because modern wood stoves are sized to burn this length of wood. Leftovers from the cutting of wood can make up the entirety of your walls, if you are willing to re-cut the wood, and this would be the most eco-friendly method of building a cordwood home. However, you might end up with walls one foot thick, which means less thermal efficiency. Alternatively, you can use precut cordwood.

Though cement is the typical mortar, you can also "cob" the lengths of wood as described previously, providing you have a post-and-beam load-bearing structure. The use of cob is more environmentally friendly, as Portland cement emits toxins during mixing, though these diminish over time as the mix dries.

When using cordwood in damp climates, you must have a raised foundation, deep eaves on the roof, and some type of waterproof sealant on the outside walls to prevent the wood from absorbing moisture and deteriorating or later forming mold. Plaster or stucco would hide the rustic effect of the exposed log ends, but you can seal the exposed ends with linseed oil periodically or use an eco-sealer such as soy, mentioned previously.

For more information on cordwood homes, visit **www.greenhomebuilding.com/cordwood.htm**.

Advanced Framing Techniques to Eliminate Lumber

Advanced framing techniques, sometimes called Optimum Value Engineering (OVE), is a house-building technique designed to reduce the amount of lumber used.

Stick-built, or traditional houses, use an inordinate amount of lumber in the form of 2-by-4s, or 2-by-6s, to reinforce the walls. These 16-inch-on-center boards are referred to as studs and can account for more than 25 percent of a house's lumber requirements.

Using prestressed insulation board, called structural insulation panels (SIPs), to replace some of this wood is a sustainable building practice that cuts construction costs and increases the house's insulation value. Setting studs at 24 inches rather than 16 can also save lumber, as can certain newly developed corner techniques. Advanced framing techniques can save the homeowner up to $1,000 in materials alone and between 3 and 5 percent of labor costs. Similar to post-and-beam but not as expensive, these techniques also include insulated-header options, either by cutting pre-stressed insulation or buying it ready-made. Christine and Sam Simonetta used SIPs in building their new home, and you can find their Case Study under the section on wind turbines, another environmentally friendly innovation in the Simonetta household.

Unfortunately, these techniques are still in their infancy, so it might be difficult to find a contractor in your area who has enough experience to tackle the project, even though OVE is Leadership in Energy and Environment Design- (LEED) approved. Additionally, houses designed to be stucco-sided might not be able to use the 24-inch-on-center framing technique, because cracks may develop in the stucco. For more information on Optimum Value Engineering techniques, visit **www.eere.energy.gov** and type in OVE.

Insulation

Before you insulate your home, you need to consider where the insulation is going and what kinds of R-values are mandated by local building codes. In my area, codes recommend R-18 wall insulation, R-49 attic insulation, and R-25 floor insulation. To find recommended R-values for your area, visit **www1.eere.energy.gov/consumer/tips/insulation.html#map**.

You can choose from a variety of insulating materials, which can be divided into four main categories: rolls, or batts; loose fill; rigid foam; and foam-in-place applications. You can choose from natural fibers or chemical composites. Rolls include fiberglass or rock wool, available in standard widths to fit 16-inch on-center stud wall formations. These rolls can be four or six inches thick, depending on your house's framing, and provide values from R-13 to R-21, depending on their thickness and content. Fiberglass and rock wool have equal R-values in roll form, but rock wool has more density when used as a loose fill. Both might cause skin reactions, and fiberglass in particular contains tiny fibers that can become embedded and break down, releasing formaldehydes. Rigid foam and foam-in-place insulations may be polyurethane, polystyrene, polyisocyanurate, and phenolic open-or-closed cell foams, though there are environmentally friendly alternatives as well.

Natural Fibers

Natural-fiber insulations include cotton, wool, hemp, and straw. Bonded Logic (**www.bondedlogic.com**) offers a cotton insulation made from 85 percent postindustrial recycled natural fibers such as denim. It is a Class A building material that meets both Uniform Building Code standards and ASTM requirements for both commercial and residential batt insulation. It comes slightly wider than standard batts, to ensure a tight fit and fill capacity, and uses only boron as a fire retardant, which also helps control the growth of fungus and resists pests. UltraTouch has excellent sound-dampening qualities, requires considerably less energy to manufacture than foam insulation, has an R-value of about 3.4 per inch of thickness, comes in depths from 3½ inches to 8 inches, and has a one-hour fire rating. Cotton insulation may cost about 15 percent more than standard fiberglass batt insulation, however the environmental and indoor-air-quality benefits are worth it.

You can also use sheep's wool for insulation. It is, like cotton insulation, treated with boron as a fire-and-pest retardant, and repeated wetting might leach the borates, though there appear to be no ill effects as a result. Sheep's wool has the same R-value as cotton, or about 3.4 per inch. In Germany, wool insulation is being used to remediate "sick building" syndrome, as a result of research that demonstrates sheep's wool can absorb and permanently retain high levels of formaldehyde, nitrogen dioxide, and sulfur dioxide. A two-inch thick, 24-foot long batt of wool will cost about $6 per pound, or about $100 per batt, making it very expensive but worth the cost for its air-remediation properties.

Hemp is not available as an insulation material in the United States, because both hemp and marijuana (cannabis sativa) are illegal to grow. Hemp is the "industrial" version of marijuana, and lacks the cannabinoids, or THC, that make marijuana a psychotropic drug. In the Netherlands and Germany, hemp is grown and made into a natural-fiber insulation and building material and sometimes woven into fabric.

Cellulose, or Loose-fill Insulation

Loose fills are particles of an insulative material blown into a cavity where it would otherwise be hard to install roll insulation and include fiberglass, rock wool, and cellulose. In this category, dense-pack cellulose offers the best R-value and the most eco-friendly aspect. Made from 80 percent post-consumer recycled newspaper, it is also environmentally friendly and requires less energy to manufacture than fiberglass. The downside of cellulose is it is both hydrophilic and hygroscopic; that is, it attracts and holds water. If you have a leak, your cellulose insulation can turn into a "wet blanket," providing no insulation value and turning your insulation into a corrosive. The chemicals used as fire-retardants in cellulose can, when wetted over extended periods of time, cause metal fasteners, metal plumbing, and electrical wires to corrode. Because cellulose is dusty, it can also cause respiratory distress when installing, if proper protection is not used, and this effect can extend over a long time if walls or barriers have gaps through which the cellulose dust can pass.

Foam Insulations

There are two types of foam insulation: rigid sheets and sprayed-in-place applications. Rigid foam insulation comes in sheets, 4 feet by 8 feet, in a number of substances, including polyurethane, polystyrene, polyisocyanurate, and phenolic compounds. Polyurethane, which comes in sheets and as spray-on insulation, has an R-value of between 5.6 and 7 per inch of thickness. Polyurethanes should not be confused with urethanes, or ethyl carbamate. Unfortunately, polyurethane foam does contain fire retardants, which can be either brominated retardants, such as penta-brominated diphenyl ethers, penta-BDEs, or phosphorus-based retardants. No mandatory labeling exists to identify which retardant is used, and penta-brominated diphenyl ethers pose a recently discovered and rising health concern. In studies on animals, these chemicals damage the liver, affect developing brains, and lower thyroid hormone production. Levels of penta-brominated diphenyl ethers in human breast milk are now 15 times higher than they were in 1997, and they have been voluntarily banned in Europe.

Polystyrene foam comes as expanded polystyrene (EPS), with an R-value of about 4 per inch of thickness. Extruded polystyrene has a slightly higher value of 5.

Polystyrene is a petroleum by-product. It is no longer manufactured with chlorofluorocarbons as blowing agents but still represents significant health risks. Polystyrene uses benzene in its manufacture, and benzene is a Class 1 substance on Material Safety Data Sheets and is known to cause cancer. Extruded polystyrene insulation, commonly light blue or pink, is extremely flammable and produces large amounts of toxic, black smoke. Also, polystyrene does not degrade but breaks down in the presence of sunlight into increasingly smaller particles, which are ingested at all levels of the food chain. A 2007 report published in the American Journal of Epidemiology called Exposure to Styrene and Mortality from Nervous System Diseases and Mental Disorders states, "Long term exposure to small quantities of styrene can cause neurotoxic symptoms like fatigue, nervousness, and difficulty sleeping, low blood platelet and hemoglobin values, chromosomal and lymphatic abnormalities, and carcinogenic effects."

Polyisocyanurate rigid foam has the same R-value as polyurethane foam and contains the same fire-retardant chemicals, again unlabeled. Polyisocyanurate foam, like polyurethane foam, will produce carbon monoxide when burning. Polyisocyanurate foam board also comes in different strengths, and its application can increase the structural stability of a dwelling.

Next to phenolic closed-cell foam, which has an R-value of 8, polyurethane is the next best spray-in insulator, with an R-value of 5.6 to 6.2 per inch of depth. However, phenolic foam, when wet, can accelerate the corrosion of metal components such as joist hangers, nails, and screws and has been the subject of litigation for its degradation effects. Phenolic foam also shrinks after being applied, sometimes as much as 2 percent.

Foam's single virtue is that its R-value is consistently twice that of other insulating materials of the same thickness.

Environmentally Friendly Foams

There are environmentally friendly alternatives to all the above-mentioned insulators. One is Air Krete, a nontoxic spray made of air, water, magnesium oxide, and ceramic talc and has the uncured consistency of shaving cream and superior insulative properties, measuring R-3.9 per inch of depth. Air Krete does not require massive machinery to install and creates no dust. Unfortunately, like most newly developed products, which lack a wide consumer base, it also costs more. To learn more, visit: **www.airkretecanada.com/home1.htm**.

Alternatively, you can use the new generation of polyols, based on naturally occurring vegetable oils such as corn, soy, and castor oil. Manufactured by such companies as BioBased Insulation, one particular item — marketed under the trade name Agrol — is both eco-friendly and Uniform Building Code standard-approved and offers thermal values of R-3.8 per inch of depth. Agrol comes in both open-cell and closed-cell foams, which expand to 100 times their size within seconds and will not shrink, settle, or disintegrate. Both open- and closed-cell foams have a Class 1 fire rating as well. More important, according the manufacturer, is the fact that every pound of product that replaces a petroleum-based equivalent provides a nearly 6-pound benefit to the environment, and these foams are water-blown, not dispersed with hydrochlorofluorocarbons. For more information, visit **www.biobasedtechnologies.com**, or read the next Case Study, provided by Coler Insulation.

CASE STUDY: COLER NATURAL INSULATION

Coler Natural Insulation: Residential Insulation Project

By Coler Natural Insulation, Ionia, New York

Kathy Coler, President (kathy@coler.com)

Project Summary

Project Opportunity: The owner of a 1926 "Sears Catalog" home in Rushville, New York, sought to remediate unexpected problems associated with the installation of a new coal-burning stove during February 2005 and to mitigate heat loss, particularly on the second floor of the home. Her primary concerns were significant ice-damming, inability to keep the upstairs warm, and associated high propane supplementary fuel costs to maintain a temperature above freezing. These problems were caused by no insulation in the attic space, which led to heat loss through the roof.

Project Goals: The homeowner's primary goals in insulating her attic were to prevent ice-damming and the associated roof damage, establish a comfortable upstairs living space, and to decrease household heating costs. Coler Natural Insulation also viewed the project as an opportunity to demonstrate the energy-effectiveness of the environmentally preferred BioBased brand of spray foam insulation.

Project Activity: The insulation product, BioBased 501, a soy-based, spray-on

CASE STUDY: COLER NATURAL INSULATION

expanding foam, was applied to several areas of the house that had previously lacked insulation.

Project Benefits:

- Enabled entire residence to be maintained at a comfortable, constant 65°F or higher during the winter
- Ice-damming eliminated
- Energy costs reduced significantly
- Soy-based product uses approximately 30 percent less petroleum resources than traditional polyurethane foam products.

Environmental Benefits

Indoor Air Quality

- Product does not release any formaldehyde or other volatile organic compounds (VOCs) or particulates during or after application.

Reduction of toxic or persistent chemicals

- Product does not contain hydrofluorocarbons (HFCs) or formaldehyde.

Renewable Resource Use/Environmentally Preferable Product Procurement

- Soy-based foam insulation product uses approximately 30 percent less petroleum resources than traditional polyurethane products.

Energy Efficiency

Project led to an increased consumption of energy due to removal of cost prohibition and increasing the ambient internal upstairs temperature from 40° to 65°F.

Financial Benefits

Cost Savings

- Savings of approximately $120/month were achieved on energy-related bills since coal is significantly cheaper than propane.

Payback Period

- The insulation project will pay for itself in about four years.

CASE STUDY: COLER NATURAL INSULATION

Other Benefits

Miscellaneous

- No further ice-damming occurred after insulation was installed.
- Entire residence is now able to be maintained at a comfortable constant 65°F during the winter.

PROCESS CHANGES

Before	After
• Insufficient insulation in place	• BioBased 501, a soy-based, spray-on, open-cell, expanding polyurethane foam, was applied by a trained and certified specialist to the underside of the roof (a Cathedral-style attic ceiling), hot water and heating pipes, rim boards around the perimeter of the basement, and parts of the window frames. Weather stripping was also installed around the windows to minimize drafts.
• Uncomfortable temperature of 40°-50°F upstairs. Inability to maintain a constant temperature upstairs.	• Inside temperature throughout the entire house is now maintained at a constant 65°F during the winter.
• Ice damming occurred regularly, putting a new roof in jeopardy.	•No ice damming since insulation was installed.

Barriers/Challenges/Regulatory Issues

Some code officials are unaware of the effectiveness (high R-values and air sealing capabilities) of foam insulation as compared to fiberglass. This lack of credible information sometimes deters clients from choosing a natural insulation product over a typical type, such as fiberglass.

Awards/Environmental Programs

The BioBased company was awarded the "Green Building Product of the Year" by Greenbuild in 2003. Users of BioBased insulation might be eligible for the US Green Building Council's LEED credits in the following categories:

CASE STUDY: COLER NATURAL INSULATION

- Energy & Atmosphere
- Materials & Resources
- Indoor Environmental Quality
- Innovation & Design Process

Applicability to Your Business

Property Owners/Developers

Proper insulation (and ventilation) of attics and roof spaces can considerably reduce both energy consumption and the potential for ice damming that occurs during the cold winter months, which can lead to other problems, such as water damage.

Lessons Learned

RGBN Program Director Interpretation

Although the quantity of energy used increased, the project was a success for the homeowner since it enabled the new coal-fired heating system to operate at its most efficient and the household to be maintained at a comfortable and safe temperature throughout the winter. It also eliminated ice damming while reducing energy costs.

Use of renewable resources to produce chemical products, such as insulation foam, is becoming more prevalent as we move away from our petroleum-based economy. This path is not without its problems, as it will inevitably lead to competition for land between food, chemical, and energy production.

Environmental Background Info

Energy Consumption in Buildings

Nearly 40 percent of all primary energy in the United States is consumed in buildings, and of that, about half is directly consumed in space heating. (Source: U.S. Department of Energy.) Inadequate insulation and air leakage are two of the primary causes of wasted energy in U.S. homes, and therefore, proper insulation of residential buildings has significant potential to reduce primary energy consumption in the United States.

The benefits of foam insulation include the product's ability to expand to create airtight seals, reducing air leakage and heat loss. The indoor environmental benefits of spray-foam insulation include its inert chemical structure, which prevents off-gassing.

CASE STUDY: COLER NATURAL INSULATION

Renewable Resource versus Petroleum Based

Of the existing spray-foam polyurethane insulation products, most are produced from petroleum derivatives, a non-renewable resource. Soy-based insulation products are derived from soy, a renewable resource. BioBased spray-foam insulation is derived entirely from soy grown within the United States.

Coler Natural Insulation is a family owned contractor specializing in environmentally preferred insulation products for residential and commercial buildings. The company can help builders, consumers, and communities find the right mix of construction solutions to sustain and preserve our environment. It is located at **www.coler.com**.

A third alternative is Icynene. Though an isocyanates compound, it uses mostly water to achieve the fill effect and uses carbon dioxide as a propellant, rather than chlorofluorocarbons or hydrochlorofluorocarbons (HCFCs). HCFCs are currently being used to phase out ozone-depleting chlorofluorocarbons. Unfortunately, HCFCs also deplete the ozone, albeit to a lesser degree than chlorofluorocarbons. HCFCs are considered an intermediate step on the path to a fluorocarbon-free economy and will eventually be replaced by eco-friendly propellants, such as carbon dioxide. In the meantime, HCFC-141b, used in blowing foam insulation, and scheduled for removal in 2004 under EPA rules, has a global warming potential 1,800 times greater than that of carbon dioxide. The only reason industry continues to use the more costly fluorocarbon approach is that the technology and infrastructure exists to accommodate them, and industry is unwilling to invest huge amounts in retrofitting to use other methods. You can sway them by refusing to purchase their products until they mend their ways. The fact that it will cost you slightly more should not sway you if you want an earth habitable to future generations. In that respect, Icynene is ideal.

Heating and Cooling Systems

You can choose a traditional, gas-forced air heating and cooling system, or HVAC system, with Energy Star approved units. If you have hot water baseboard heat, or radiators, you can choose Energy Star boilers, which have a higher annual fuel utilization efficiency rating, which is another measure of heating equipment efficiency. To further increase efficiency in both forced-air and heated-water systems, you can choose electronic ignitions, where the pilot light is intermittent rather than constant.

You can add a programmable thermostat, preferably an Energy Star-approved model, which contains no mercury, and set it so that your house is not overly warm when parents are working and children are at school, or at night when everyone is bundled under blankets. Appropriate use of programmable thermostats can save the average family about $150 a year. Install your programmable thermostat on an inside wall, away from heating or cooling ducts, drafts, or windows whose sunlight and consequent heat may affect the thermostat. If you have a ground or air source heat pump, you might require a special thermostat.

Before you consider replacing your heating and cooling system, you can perform a number of upgrades to improve its efficiency. You can weatherproof your home by sealing windows and doors and other areas where outside air leaks inside. You can insulate walls, attics or ceilings, foundations, and crawl spaces. The most effective use of insulation is in an attic or ceiling because heat rises and escapes upward.

You can also inspect and upgrade your ductwork, if you have a forced-air furnace. You can check for leaks with diagnostic equipment or hire a professional. You should replace all damaged ducts, reconnect ducts that have become disconnected, and replace undersized ductwork. You can also straighten out the ductwork, make sure each duct meets and is sealed to its corresponding register, and seal the joined areas with metal tape or a sealant; duct tape, in spite of its name, is not an enduring seal. You can also insulate around ductwork with specially formulated insulation in attics and crawl spaces. Afterward, complete your upgrade by replacing your furnace filter.

After you upgrade your ductwork and add insulation, your house will be more airtight. This restricted flow of air from outside means your furnace, gas dryer, and gas hot water heater are using more of your house's available oxygen, and this can be dangerous. It is important to perform a combustion safety test on both your heating unit and other appliances. You may need to add some form of ventilation. You might also create a carbon monoxide environment if your heating equipment is not operating properly. You should, in any event, have both carbon monoxide and radon detectors inside your home.

If you choose to have a contractor do the work of upgrading, make sure you receive a detailed bid outlining the parts to be replaced, preferably by model number or manufacturer, as well as the costs and the time frame in which the upgrades will be performed. Contractors who work on your heating and cooling system must have a certificate for refrigerant handling, contractor's liability insurance, and a state license. They should also have several years of experience in business and would be able to provide customer references on request. Always check with the Better Business Bureau.

You need to upgrade your equipment or insulate your home and upgrade your ductwork if some of your rooms are too hot or too cold or if your home has humidity problems, which can be identified by condensation on the windows or mold buildup. If you home is excessively dusty, or your heating or cooling system is very noisy, you may also need to upgrade or retrofit. Furnaces, boilers, and air conditioning units have an effective life of about 15 years. If you buy new systems, always select Energy Star-approved systems, keeping in mind that bigger is not better, and oversized systems will cycle more frequently, costing you extra on your utility bills. Energy Star systems are more expensive but will lower the cost of your utility bills, pay for themselves over their lifetime, and add to the resale value of your home. Many come with rebates provided through your local utility. You might also qualify for a federal tax rebate. To find out about rebates, visit the Energy Star Web site at **www.energystar.gov/index.cfm?fuseaction=rebate.rebate_locator**.

Size Matters

Proper sizing of your heating and cooling system is essential to proper functioning. Because standards for sizing were not universally available until recently and most homes were severely under-insulated, many heating and cooling contractors sold their customers oversized systems, resulting in rooms that were colder, or warmer, than the rest of the home and heating or cooling equipment that needed frequent repair.

Even now, some contractors are unaware that sizing standards exist and so install systems larger than needed by the size or configuration of the dwelling. If you are considering replacing your system, be sure your contractor is aware of these universal specifications for sizing, and do not allow him to "match" you new system to the old by simply replacing one 50,000 Btu unit with another. Neither is sizing your system according to the square footage of your house or using a chart based on "averages" the appropriate way to determine the size of your heating or cooling unit, though these methods can be used to first estimate to determine costs and the amount of labor involved in replacement.

Manual J, "Residential Load Calculation," published by the Air Conditioning Contractors of America (ACCA) is the approved method for sizing heating and air conditioning units on homes built in the United States. This manual takes into consideration a number of factors, including climate, insulation levels, air infiltration rates, the types of windows and their location and size, the configuration and orientation of the house itself, and even the number and ages of the occupants and their comfort preferences. It also takes into account lighting and appliances, both of which give off heat. The manual is difficult to use, but there are also worksheets and even software packages that can help your contractor determine the appropriate size of your heating or cooling unit, and you should insist that he or she use at least one of them. If your contractor is not familiar with this manual, your local utility might be able to help.

Where ductwork is part of the installation, there is another manual, called Manual D, or "Residential Duct Design," and also produced by the ACCA, which offers specifications for duct design and installation. The ACCA also offers Manual S, or the "Residential Equipment Selection" manual, a comprehensive guide to selecting residential heating and cooling systems. With all this information freely available, either on the Web or through your local utility or a reputable heating and cooling contractor, there is no reason you should be stuck with an oversized, inefficient, and costly system.

The one exception to the above rules and manuals is a steam-heating system. Here, the boiler should be sized to match the size of the radiators. Even then, with insulation and window upgrades, it might be possible to remove a number of radiators, or replace them with smaller units, and still maintain comfortably warm rooms. If you replace radiators, you will probably have to replace the shutoff valves in a correspondingly smaller size. Visit a recycling center, an antique salvage yard, or Internet sites such as CraigsList or eBay — to find the needed replacements. Old radiators are readily available. If you do not find what you need, you can also buy them new. Unfortunately, the newer models that I found are both very pricey and lacking in the ornate detail of antique radiators.

When the contractor finishes sizing your system, using the specifications contained in the above manuals, ask for a copy of the worksheet or a printout from the software application. This is the only way to verify the calculations were correctly made. Also ask for a written contract specifying the model number and manufacturer of the needed unit, its energy-efficiency rating, the warranty, any available rebates or tax incentives applicable to its purchase, and the proposed method of arbitration if you, the homeowner, are not satisfied with either the product or its method of installation.

To find which heating and cooling units qualify for the Energy Star label, visit **www.energystar.gov/index.cfm?c=bulk_purchasing.bus_purchasing#res_app** and follow the residential appliance link.

Energy Efficient Units

Always ask for, and purchase, energy-efficient heating and cooling units. Central air conditioning energy efficiency is rated in seasonal energy efficiency ratios (SEERs); the higher the rating, the more efficient the unit. Window air conditioning units are measured in simple energy efficiency ratios (EERs). The coefficient of performance (COP) is used to measure heat pumps, and here again, the higher the value, the more efficient the system. Heating seasonal performance factors (HSPFs) also measure the relative efficiency of heat pumps. The annual fuel utilization efficiency standard measures how efficiently gas furnaces or boilers operate, and the higher the value, the more efficient the system. Most gas furnaces and boilers on the market today have annual fuel utilization efficiency ratings of 92 percent or greater, so do not settle for anything less, except in a furnace with a continuous pilot light that generates a higher annual fuel utilization efficiency rating. In areas with frequent outages, some homeowners choose continuous pilot lights because electronic-ignition, gas-fired furnaces will not run when the power is out.

For more information on determining energy efficiency using the above-mentioned ratings, visit **www.cwlp.com/Energy_services/efficiency_ratings.htm#AFUE**.

Zoned Systems

Zoned systems work by providing different levels of heating or cooling to different rooms, depending on their use, normal hours of occupation, and the preferences of the individuals using them. In a typical zoned system, sensors monitor the temperature of a room, or area, and send signals to a preprogrammed master controller to alter that temperature as needed. If you are in your bedroom only at night, there is not much point to heating it to 75°F all day; conversely, if you are in the kitchen most of the day and some of the evening, there is no reason to suffer cold feet because your thermostat is set at 70°F and you have tile or stone floors.

New technology, incorporated into such devices as flat panels and 1-2-3 wiring configurations, makes zoned systems more economical and much easier to install. Zoned systems are particularly useful in homes that have multiple levels, extensive length-to-width profiles or wings, many windows, or large, open areas. If specific ducts run to specific areas of the house, and if the ducts are accessible, installation is simplified. However, due to the complexity of the monitors, wiring, and the needed retrofitting of ductwork, it is best to have a professional install your zoned system.

Passive Heating and Cooling

Passive heating and cooling does not make use of any device, whether a furnace, air conditioner, or even a solar panel, but instead relies on natural energy flows and sources to maintain comfortable temperatures. In order for a home to take advantage of these natural energy flows, it must be built and oriented with features relative to the climate and be sited to take advantage of them. Incorporating the principles of passive design into your new existing home will make it more comfortable and energy efficient.

Passive solar heating is first approached from a design standpoint; passive heating plans commonly include a south-facing wall, often with windows or another sunlight-absorptive surface, and are often canted to match the angle of the winter sun and possibly bermed to provide a thermal mass, which will also absorb the heat of sunlight and transfer it to the inside.

Passive heating uses rectangular floor plans, elongated on an east-west axis, and overhangs of sufficient depth to deflect summer sun while still allowing winter sun. Passive heating systems rule out windows on the north side of a house in the North and allocate space use based on how often an area is used and how warm it needs to be to please the occupants. Areas such as closets, storage rooms, and garages are oriented along a north wall, where they can act as a temperature buffer to the rest of the house. Entryways, traditional sources of heat loss, are oriented to the south and commonly include vestibules, or porches, which can mediate the air temperature coming into the home. The addition of a southerly oriented greenhouse further absorbs and distributes heat. Roof ponds also incorporate passive solar techniques.

In a passive design, all heat flow is accomplished via radiation, convection, and conductance. This does not mean that no mechanical assistance is provided to distribute heat, and ceiling fans are often used for this purpose. Passive-assisted systems may be referred to as "hybrid" heating systems. The Rural Renewable Energy Alliance provided some hybrid systems in Minnesota, and Rin Porter wrote an excellent study for the Clean Energy Resource Teams detailing this program.

CASE STUDY: RIN PORTER

Solar Space heating System Warms Rooms for Sebeka Family

By Rin Porter, Central CERT

A Seboka family has become the first in their neighborhood to have a solar space heating system installed in their home. The solar space heating system will heat two bedrooms with energy from the sun. It does not use electricity or any other energy source for its operation.

The solar space heating system was installed through the Minnesota state fuel assistance program funded by the federal government. The fuel assistance program provides money to pay part of a family's heating bills during the cold Minnesota winter.

CASE STUDY: RIN PORTER

Once a family has qualified for fuel assistance for the cold weather months, they also become eligible for the solar assistance program, which is administered by the Rural Renewable Energy Alliance (RREAL), a 501(c)(3) nonprofit organization whose primary purpose is to bring solar heat to lower-income households.

While fuel assistance is a valuable aid to low-income families to help them stay warm in the winter, it does nothing to deal with what RREAL sees as the real problem: Dependence on expensive fossil fuels. A family may receive fuel assistance for many years, and that money is simply burned up with the coal, natural gas, or propane that the family is using to heat their home.

But if the family changes over to a renewable energy heating system, it will no longer need fuel assistance. That is where the Rural Renewable Energy Alliance saw the opportunity to make a difference for many low-income families: provide them with a new heating system that uses the sun, not coal, natural gas, or propane.

RREAL installs solar energy-based technology as an alternative to coal, natural gas, or propane fuels because solar energy is renewable and nonpolluting. Renewable energy sources include wind, biomass, and water, along with solar. Yet, renewable energy remains out of reach of many lower income households due to financial and informational barriers.

Here is how the solar space heating system works. First, Jason Edens, director of RREAL, visits the home where the system is to be installed to find the best location for the solar panel to be put up. The panel must be installed on a south facing outside wall. The panel will absorb heat from the sun and produce warm air. Flexible tubing will bring the warm air from the solar panel to a room inside the house.

Next, Edens examines the inside wall of the room or rooms to be heated, to plan how to run the tubing from the outside of the house to the room or rooms. A hole is cut through the exterior wall siding and interior wall covering to allow the tubing to enter the home. The tubing is connected to the solar panel, and the solar panel is installed on the exterior of the home.

In the Sebeka home, Edens ran the tubing into the house through a bedroom closet. Using a Y-connector, tubes were run through that closet and another one in the adjacent room to provide space heating to two bedrooms. Both rooms will be heated by the solar space-heating panel on the outside of the home.

"We are thrilled about this," said Mary, one of the homeowners. "This solar heating system is completely self-contained and not electric at all. If the power goes off, this s system will work, and at least during the day we'll be warm."

CASE STUDY: RIN PORTER

Mary learned about solar energy from her parents, who have a solar electric system to power part of their home. She heard about the RREAL solar assistance program from her dad.

"This is the first year we've had to apply for fuel assistance," Mary said. "We've lived in our home for five years, and we managed to pay for everything up to now. But this has been a difficult year for us. My husband was out of work for six months, but now he's working again part-time."

Mary's home uses a natural gas furnace as the main source of heat. The furnace heats the main rooms of the house well, but leaves two of the bedrooms cold. These two rooms will be heated by the new solar space heater.

Connie Warner, energy assistance coordinator with Otter Tail-Wadena Community Action Council in New York Mills, which operates the Otter Tail and Wadena County fuel assistance programs for the state, said in a telephone interview that during the 2004-2005 heating season, 944 families in Wadena County received fuel assistance. So far this fall, 795 families have already applied for fuel assistance for the 2005-2006 heating season.

Fuel assistance is based on family income for the ninety days preceding their application. For a family of one, the income limit is $4,987. For a family of two, it's $6,527. For three, it's $8,056. For four, it's $9,591.

The RREAL solar space heating installation in Sebeka is the first one in Wadena County. Systems have been installed in Cass County, Crow Wing County, and the White Earth reservation. Up to 15 other installations are planned for the next 12 months.

As often as possible, RREAL involves local "at-risk" youth referred by Community Concern for Youth or other agencies in the solar installations, teaching them how to assist in the process. Involvement in meaningful service projects like this has led some of these young people to decide to complete their high school education, take more science classes, and continue to volunteer in their communities.

RREAL is funded by donations from individuals and grants from the University of Minnesota Central Region Partnership, the Rural Poverty Fund, the Initiative Foundation, Otter Tail-Wadena Community Action Council, and White Earth Land Recovery Project. If you would like more information about the solar assistance program, or would like to purchase a system, contact RREAL at 218-587-4753, or visit their Web site at www.rreal.org.

CASE STUDY: RIN PORTER

Project Snapshot:

Technology: Solar Space heating system

Benefits:

- Family will no longer need fuel assistance with renewable heating system
- RREAL places renewable energy within the reach of low income families
- Involvement of local youth

Case Study Courtesy the Rural Renewable Energy Alliance, Minnesota and Joel Haskard, Co-Coordinator, Clean Energy Resource Teams (CERTs) Regional Sustainable Development Partnerships

Passive heating and cooling systems also take advantage of landscaping features, like trees and shrubs to deflect the cold north wind, or natural bluffs or rock outcrops to shelter the entrance to a passive solar home. An adequate description of all the features of passive solar design, or passive heating and cooling, would require another book-length explanation. Newer hybrid systems even combine active and passive solar, and you can find out more about this revolution at **www.radiantsolar.com/rodales.pdf.**

To find our more about passive solar technology and how you can use these techniques, either in your existing home or when building a new home, visit: **www.azsolarcenter.com/technology/pas-2.html**, or review the case study below, provided by Okamato Saijo Architects.

CASE STUDY: PASSIVE HEATING & COOLING

By journalist Laurinda "Rin" Porter, The Rural Renewable Energy Alliance, RREAL, Minnesota, and Joel Haskard, Co-Coordinator, Clean Energy Resource Teams (CERTs), Regional Sustainable Development Partnerships

Passive heating and cooling systems also take advantage of landscaping features, such as trees and shrubs, to deflect the cold north wind, or natural bluffs or rock outcrops to shelter the entrance to a passive solar home.

CASE STUDY: PASSIVE HEATING & COOLING

An adequate description of all the features of passive solar design, or passive heating and cooling, would require another book-length explanation. Newer hybrid systems even combine active and passive solar, and you can find out more about this revolution at **www.radiantsolar.com/rodales.pdf.**

To find our more about passive solar technology and how you can use these techniques, either in your existing home or when building a new home, visit: **www.azsolarcenter. com/technology/pas-2.html,** or review the Case Study below, provided by Okamato Saijo Architects.

Sonoma Home Bridges: Modern Living with an Environmental Ethic

The Johnson-Theis Residence, Sonoma County, California

Carolyn Johnson and Rick Theis wanted a private oasis and a place to host large parties. They also wanted their new home to reflect their lifelong commitment to conservation and restoration. They needed separate home offices for their work, guest rooms for their family and friends, and plenty of storage space, including a wine cellar for tastings. Their new home, designed by Paul Okamoto of Okamoto Saijo Architecture, meets these myriad demands.

The design problem was to create this oasis on an undistinguished site — an old Gravenstein apple orchard, abandoned 60 years ago, bordering a forest of second-growth redwood and fir trees. One solution was to locate a sun terrace on the roof and a home office inside a three-story tower that captures the dramatic view of the Sonoma Valley and Santa Rosa Plain, with vineyards and forests in the foreground and Mount Diablo in the distance. Another solution was to span part of the library building wing over a new dry creek bed with boulders and drought-resistant plants. During the rainy season, this creek bed acts as an important water drainage route. A narrow, 40-foot long pedestrian wood and steel bridge also spans this creek bed and leads to a covered porch over the front door.

The home pivots around the 825-square foot, wood-trussed great room, designed for dinner parties of 20 to 30. The kitchen cabinets were designed by Johnny Grey. The great room is elevated to face the sweeping views of the Green Valley to the southeast and is aligned along a curved line for solar orientation. The north wall of the great room faces the uphill terrace for outdoor summer parties and the enclosing view of the surrounding forest to the north. Adjacent to the great room is a pantry/prep kitchen built with sprayed earth walls to embody the essence of an Old World kitchen.

The building captures the sun through the south windows during cool winter days and avoids it during the hot summer. Solar panels on the roof produce electricity, thereby reducing the demand on the electrical grid. Additional solar panels provide hot water.

CASE STUDY: PASSIVE HEATING & COOLING

Concrete floors and earth walls act as thermal mass, which helps to maintain a constant room temperature throughout the year.

The home showcases more than a dozen sustainable building materials, including lumber certified by the Forest Stewardship Council as sustainably harvested and old timbers that are reused as exposed beams for the house. The redwood staves used for exterior siding came from old water tanks originally milled in Mendocino County, probably by Carolyn's great-grandfather. Concrete used in the foundation contains 30 percent fly ash, an industrial waste. The material for the sprayed earth walls came from Nuns Canyon, a local quarry in Sonoma Valley.

Wood Burning

Burning wood, in a stove or fireplace, is a wonderful way to create a warm ambience in your dwelling. It is not, however, environmentally friendly, unless you have a wood lot you regularly replenish by planting trees and a wood stove or fireplace equipped with a catalytic converter.

Catalytic converters for wood stoves and fireplaces are very similar to the smog-control devices on automobiles and allow the volatile gases to burn at lower temperatures. The converter itself is a honeycomb-shaped baffle coated with a catalyst, which allows complete smoke combustion and heat release at a minimum temperature of 500 degrees. Catalytic converters need to be replaced after about five years because they lose their efficiency.

Non-catalytic stoves also use baffles, and sometimes secondary combustion chambers, which route burnable gases through the hottest part of the firebox and mix them with air to burn them more thoroughly. Catalytic stoves can achieve up to four stages of combustion before what remains of the particulate in the smoke is released from the chimney.

If you plan to burn wood, either in a stove or fireplace, choose an EPA-certified stove. These stoves meet a particulate emissions standard of 7.5 grams per hour for non-catalytic stoves, or 4.1 grams per hour for catalytic stoves; standards are even stricter in the state of Washington and mandate either 4.5 grams or 2.1 grams, depending on whether the stove is equipped with a catalytic converter. The Clean Air Act mandates these standards for newly manufactured wood stoves, and you can identify their EPA certification either from a paper label attached to the front or a metal label affixed to the back of the stove. You can find a list of these stoves at **www.epa.gov/compliance/resources/publications/monitoring/caa/woodstoves/certifiedwood.pdf**.

In California, smoke from wood-burning stoves has become a health hazard, particularly in the Sierras, where such towns as Quincy and Greenville regularly suffer smog in the winter due to accumulations from wood-burning stoves. I lived in the area for eight years and often refused to travel "down the mountain" on winter days because I have asthma and could not tolerate the levels of air pollution. Reno, 45 miles to the southwest, was not a great deal better. Burning wood creates carbon monoxide, which can cause dizziness, severe headaches, and long-term health problems. Burning wood also creates nitrogen oxide (NO_2), which further affects the respiratory system and contributes to acid rain. The volatile organic compounds contained in wood smoke include benzene, formaldehyde, and benzo-a-pyrene, a polycyclic aromatic hydrocarbon, and some manufactured logs can even contain polychlorinated biphenyls. Burning kiln-dried lumber scraps might be economical but results in significant creosote buildup, which can cause your roof to catch fire.

Tips on Fire Building

Start your fire with a soft wood, such as pine or fir. These woods ignite easily and burn fast, producing a great deal of heat to warm the inside of the stove. They also form creosote readily and so are not recommended for continuous wood burning. If you must use them to heat, burn creosote-removing logs or bricks at frequent intervals to keep your chimney free of creosote, and have it professionally cleaned at least once a year, preferably at the beginning of winter. You should also conduct yearly inspections of your chimney and chimney cap, your catalytic converter, all stove piping, including angles and fasteners, any gaskets on the doors or seams in your stove's walls, and the firebrick under and around your stove.

Hardwoods, such as oak and walnut, burn slower and more evenly and produce less smoke. They also provide more heat per mass than softwood. For more information on the heat-producing properties of various woods, see the chart below.

SPECIES	BTU/CORD
Black locust	26,500,000
Hickory	25,400,000
Beech	21,800,000
Hard maple	21,800,000
Red oak	21,700,000
Yellow birch	21,300,000
Yellow pine	20,500,000
Ash	20,000,000
Oak	19,200,000
Soft maple	19,100,000
Black cherry	18,500,000
White birch	18,200,000
Elm	17,700,000
Poplar	15,900,000
Hemlock	15,000,000
Spruce	15,000,000
Fir	13,500,000
White pine	13,300,000
Basswood	12,600,000

Burn only seasoned firewood; that is, wood that has been stacked and "cured" for at least six months. Hardwoods, which dry more slowly than soft woods, might take up to 12 months to cure completely. By definition, seasoned or cured hardwood contains less than 20 percent moisture. If you split your logs immediately on receiving them, or after cutting down the tree, your wood will dry better because wood dries from the outside in. Stack split lengths of wood first in a crosswise fashion, on pallets and away from any walls, to get good air circulation, and cover the stack against rain. A sunny location will dry wood faster, though the shade under trees will protect better from moisture. Later, when the wood is dry, you can stack it in a pyramid, again covering it to protect it from rain or snow.

If you buy your wood already split and dried, watch out for dark, cracked ends, where the cracks radiate like spokes, or very lightweight logs. Truly cured wood has a distinctive sound. Strike two pieces together; if they resonate, rather than making a dull, thudding sound, they are cured. Wood might have bark, but it should not peel or break easily, and no green should be visible beneath it.

Your initial fire should be small but hot, and you will need to open the damper on the stove wide to achieve this. Leave the damper open for about a half-hour, all the while adding softwood until you achieve burning coals. This is when your fire is producing the least amount of smoke. Open the door and start adding hardwood a few pieces at a time.

Again, leave the dampers open for about 30 minutes, which will give your fire sufficient oxygen to prevent creosote buildup. Watch the level of the fire. When the roaring combustion subsides, open the door, add a few lengths of wood, and close the damper almost completely. In a few hours, you will have a quiet, intensely hot fire. If your wood stove comes equipped with a blower, this is the time to turn it on. Add additional firewood quickly and carefully, to prevent smoke back-drafting into the room.

Pellet stoves, the new kids on the block in wood stove design, produce significant amounts of heat efficiently with less pollution than traditional wood-burning stoves. Most are exempt from certification because of their low particulate emissions, which amount to less than one gram per hour.

Pellet stoves are self-stoking, delivering pellets from a top orifice directly to the combustion chamber. The single difficulty with pellet stoves is the intermittent unavailability of fuel. Last winter, we were unable to find pellets in our area for the entire period of November through January. Normally, pellets are available both at lumber yards, home improvement stores, and sometimes feed stores.

If you burn wood in a fireplace, you can buy and install a certified fireplace insert, which will meet both air-pollution and efficiency standards. These items are available in a wide range of sizes and styles, do not diminish the thermal effect of your fireplace, and allow you to watch your wood burning without worrying about your contribution to indoor or outdoor air pollution.

Building Your Own Wood-Burning Masonry Stove

In Russia, masonry stoves are known as "Grubkas." They differ from a standard fireplace in several respects. They are larger, and the entire oven, masonry arches, and flue path must be constructed of firebrick, using fire clay mortar. They also require more material, at a greater cost, and will take more time to build. Although you can use adobe, red brick, or concrete in the shell surrounding the inner firebrick core, you are limited exclusively to firebrick on the interior. Similar stoves can be found in cold-climate countries such as Finland, Sweden, Poland, or Germany. All have a high masonry mass, which traps and releases heat gradually, and a long flow path, and they are very efficient. They require less wood to heat a room comfortably and less tending. Inferior wood such as scraps, limbs, construction lumber, and even knots can be used, though wood less than two inches in diameter provides the highest efficiency.

Long forgotten, and their method of construction almost lost to time, these Grubka stoves have recently undergone a renaissance. In 1980, an innovative adobe mason from New Mexico named Robert Proctor received a grant, funded through the New Mexico Energy Institute and sponsored by the University of New Mexico, to build eight of these stoves at senior citizens' centers around the state. The elevation of the lowest center was 3,500 feet above sea level; the highest was 7,000 feet above sea level, an altitude at which combustion takes place less effectively, due to air pressure and oxygen levels. The stoves were fitted with thermocouples to calculate their thermal performance. In spite of altitude, efficiencies for the Grubka stoves ranged from 86 percent to 89 percent, or well above what would have been expected, given the existing conditions.

Most kits, like those advertised in home building and renovating magazines, are too small to heat a large space and are expensive. Now, the Adobe Builder Web site offers a 25-page builder's manual that offers plans and thermal documentation on efficiencies.

These stoves are so thermally efficient that they burn up all pollutants and so require no catalytic converter or multiple combustion chambers. The chimney produces no smoke but only a heat mirage, indicating a probable flue temperature near 1,100 degrees Fahrenheit, according to Jay S. Jarpe, a research engineer who wrote the manual.

You can find this manual at Adobe Builder's Web site: **www.adobebuilder.com**.

Finding "Green" Architects and Contractors

The Green Building Council maintains a Web site, Leadership in Energy and Environmental Design (LEED), which lists building professionals accredited by the Council for their understanding of green building principles and their familiarity with Leadership in Energy and Environmental Design requirements. Visit the United States Green Building Council at **www.usgbc.org**.

Nonprofit Co-op America's green pages also list various architects and designers who have been pre-screened for their eco-sense. Go to **www.coopamerica.org**.

Many green architects and contractors are members of a local chapter of the Business Alliance for Local Living Economies (BALLE). You can find these members by visiting **www.livingeconomies.org**.

You can also find green architects and building contractors by searching the Green Builder's Directory online or going to an Internet site called Low Impact Living. Low Impact Living allows you to choose from architects, builders, and contractors or even look for green furnishings and apply for an energy audit, all without registering. The Green Builder's Directory requires registration but does not cost anything. If you want to buy a "green" home, go to a site called Listed Green. If you are looking for a list of green products, go to Building Green at **http://www.greenbuildersdirectory.com/** or **http://www.lowimpactliving.com**. Finally, if you are unable to find green local architects and contractors through the above resources, you can go to **YellowBook.com**, find a local architect, and quiz him on his knowledge and use of green building standards and techniques. If you are green-savvy, you can partner with an architect and/or builder, passing on your green techniques in exchange for increased cooperation and greater leeway in designing and building you green home.

You can also choose your favorite technology, such as bale homes, go online and find a builder advertising within the category. *The Last Straw*, a publication by, and for, bale home builders, has a Human Resource page that lists a number of bale-savvy builders by address and phone number. Michael Carroll and Deborah Anderson in Taos, New Mexico, design and build adobe homes. Danny Martinez, also from New Mexico, is an adobe builder who encourages people to build themselves and even teaches them how through the University of New Mexico's Continuing Education Department. Green is not just a style but a culture, and people within the Green Revolution are more than happy to reach out and help, even across specializations. Search them out and ask questions.

To obtain a home energy audit from someone other than your local utility, visit the Residential Services Energy Network (RESNET) and type in your home state (**http://www.natresnet.org/**). The RESNET also offers a wealth of other energy-related information, from cap-and-trade carbon emissions reduction programs to a blog where you can find fascinating tidbits of energy-related information, including the location of a mortgage lender for energy-efficient buildings.

Recycling Construction Materials

The rapidly growing construction and demolition (C&D) recycling market owes its impetus to the green revolution. According to the EPA, this country generated about 136 million tons of building-related debris in 1996. Of this, nearly 7 million tons were purely residential construction debris, according to The National Association of Homebuilders (NAHB), which developed a detailed methodology for determining this figure.

Almost 50 percent of this debris is wood. Another large component is cement, from foundations or basements. A good proportion is asphalt shingles. Other components are brick, masonite, ceramic tile, glass, plastic sheet film, porcelain, metal, insulation of all types, and ceiling tiles. All these materials can be recycled, either by remanufacture into other products or sale as salvage to builders and remodelers for use in homes.

One aspect of C&D recycling is demolition; another is deconstruction. Demolition recycles a wider range of materials but most often as scrap to industry because demolition is necessarily destructive of integral building components. Deconstruction, or the selective dismantling of a building, complements demolition by extracting antique or valuable building components intact for reuse by consumers before the wrecking ball or crane can demolish them. Together, the two practices reduce the amount of trash going into our landfills, reduce landfill fees for contractors, and offer a unique resale market for common items such as windows, doors, cabinets, siding, and trim, or classic, hard-to-find items such as old steam-heat radiators, marble mantelpieces, antique chandeliers, and intricately painted ceramic tiles. Even demolished, recaptured materials such as asphalt tiles can be recycled into roadbeds or gypsum wallboard into cat litter and soil additives.

Scrap lumber can be recycled into modern wood flooring; salvaged lumber can be made into a new floor.

As the demand for green construction increases, the availability, quantity, and variety of reclaimed or recycled products grows. Capitalism is a demand-driven economy; if you will buy it, they will make it. To find EPA-designated green construction and remodeling products, visit **www.epa.gov/epaoswer/non-hw/procure/index.htm**. Alternatively, you can check out the United States Green Building Council's Leadership in Energy and Environmental Design (LEED) at **www.usbgc.org**, or the National Institute of Standards and Technology's Building for Environment and Economic Sustainability (BEES) at **www.nist.gov**.

Before you start building or remodeling, contact either your local builder's association or your county's solid waste department to determine how to dispose of any waste you can not recycle through a salvage yard. If you are doing an extensive remodeling job on an older home, consider deconstruction, and contact a salvage yard to see if they will pay you for the materials and haul them away. Some salvage yards will conduct auctions of your valuable old materials such as wood floors and porcelain bath fixtures. Ask the salvage yards if they will barter some of your old furnishings for kitchen cabinets or other items you might need. Many salvagers offer building materials in very good to excellent condition, at a fraction of the cost of new.

Freecycle Network is a grassroots, global organization of 4 million members who are giving, and getting, free stuff for themselves and their communities. Begun in 2003 by Deron Beal, the organization is devoted to sustainability, as people from all walks of life band together to turn "one man's trash into another man's treasure." Visit them at **www.freecycle.org**. Also visit the ReUse People Web site, at **www.thereusepeople.org**, a nonprofit corporation committed to reducing the solid waste stream entering our landfills by salvaging usable building materials and providing them to individuals, businesses, and families, including low-income families in Mexico.

Conclusion

Thank you for reading this book. Maybe it has inspired you to live in a more sustainable fashion, if only by changing a light bulb. As I write this, the economy is in a downward spiral induced by the rising cost of oil and the crash of the housing market. Nothing is ever as simple as it seems, however, and some people are remodeling or even building new houses. If you are one of them, please build responsibly, with one eye on the future.

I have tried hard to present a fair case for sustainability. If there are errors, they are all mine. On the issue of global warming, for example, there are two points of view. One side says the current pattern of global warming is part of a natural heating and cooling cycle the earth goes through periodically. To prove their point, they cite such climate fluctuations as the Medieval Warm Period, which occurred between 800 and 1400, and the Little Ice Age of 1600 to 1850. The other side cites the "hockey stick effect," a descriptive title for a global warming graph reportedly first introduced by Jerry Mahlman, the former head of the Oceanic and Atmospheric Administration's Geophysical Fluid Dynamics Laboratory and later adopted by geoscientists Michael Mann, Ray Bradley, and Malcolm Hughes. This line graph shows that warming has increased exponentially since the beginning of the Industrial Revolution, or about 1900, resulting in a sharply rising curve, like a hockey stick.

I do not know whether the current episode of global warming is natural or man-made. I do know that there are about 91 chemicals in the earth's biosphere and in our bodies that did not even exist before World War II. I know that all the salamanders have disappeared from my neighborhood. I also know that I cannot safely eat more than once a week the fish I catch, which was not true as recently as 1960. Finally, I know that fossil fuels are a finite resource.

My thanks to the excellent editors at Atlantic Publishing, who caught my obvious errors and oversights with eagle-eyed discrimination. I also want to thank all the wonderful, "green" people who contributed their comments and Case Studies to this book and whose remarkable efforts convince me that sustainability is an achievable model. Last but not least, thanks to Angela Adams, the managing editor at Atlantic, who patiently guided my efforts toward a successful conclusion.

Glossary

Acid rain: Any form of precipitation containing sulfur and nitrogen, occurring or produced naturally or by human activities, which reacts in the atmosphere to produce acidic compounds that negatively affect plants, animals, and buildings.

Acrylic paint: Any paint containing pigments suspended in an acrylic polymer emulsion, which is water soluble when wet but water-resistant when dry.

Adobe: A brick constructed of earth, and sometimes other substances such as hay, and used in building.

Advanced framing techniques: A building technique in which lumber is more equitably designed, distributed, and reinforced to reduce its use.

AFUE: Annual Fuel Utilization Efficiency rating; a rating used to classify the efficiency of residential boilers that supply hot-water heat to baseboard heaters or radiators.

Air cleaner: An electrical, electrostatic, or mechanical device used to remove particulates from the air. Odors are particulate.

Air pollution: The addition of unwanted, sometimes toxic, particulates to indoor air. See also VOCs.

Algae: Single or multicelled aquatic plants that play a significant role in aquatic environments.

Alternating current: Electrical current with cyclical variations in direction and intensity, usually delivered in a sine wave, which represents the most efficient form of transmission. Sometimes also referred to as AC, or the opposite of direct current.

Anion light bulb: A compact fluorescent bulb with an ionizer built in, sometimes referred to as an anion-negative bulb, used to clean the air.

Anionic surfactants: Surfactants lower the surface tension of liquids, allowing detergents to work better in water. Ionic refers to the positive or negative state of ions in the chemicals used. Anionic surfactants carry a negative charge and commonly use sulfates, sulfonates, or carboxylates. Non-ionic compounds have no charge and commonly include ethylene oxide and propylene oxide or, less commonly, fatty alcohol esters.

ASHP: Air-source heat pump, which uses the ambient temperature of outside air to either heat or cool a house.

Black water: Water contaminated with untreated feces or toxic chemicals.

Btu, British thermal unit: a measure of thermal energy, or heat, based on the amount of energy needed to heat one pound of water by 1°F and representing about 1/3 watt.

C&D: Abbreviation for construction and demolition, used in connection with recycling.

Carbon dioxide: Carbon dioxide, or CO_2, is a greenhouse gas, occurring naturally as a result of mammalian respiration, fire, and volcanic activity, or unnaturally as a result of burning fossil fuels, such as coal and gasoline.

Carbon footprint (calculator): The measure of the use of fossil fuels and subsequent contribution to the greenhouse gases that cause global warming. This can be calculated based on the individual, the household, the industry, or the country.

Cast Earth construction: A method of home building in which earth, calcined gypsum, and other additives are poured on site into a frame to form the walls of a house.

Catalytic converter: A device that converts harmful emissions such as carbon monoxide, hydrocarbons, and nitrogen oxide to carbon dioxide. It is used on automobiles, wood stoves and fireplaces, and some small engines.

Certified wood: Wood certified by the Forest Stewardship Council, a nonprofit foundation, as being sustainably harvested, rather than illegally imported.

Cob construction: A method of building in which clay, sand, and straw or other fibrous materials are blended with water and traditionally mixed by hand, or by foot, and then placed in lumps, called "cobs," to create a wall.

Compost: Rotted vegetable matter in which beneficial soil bacteria have been allowed to proliferate, producing quantities of nitrogen, phosphorus, and potassium. Not to be confused with mulch.

Convection oven: Unlike regular ovens,

which allow for the random distribution of heated air, convection ovens use fans to circulate air and create a uniform temperature.

COP: Coefficient of performance, a measure of the efficiency of a heat pump based on the heat provided compared to the energy used to create it.

Cordwood homes: A method of building involving short lengths of wood, usually unpeeled, in a post-and-beam construct, anchored in place with cement, cob, or other mortars. Also, cordwood is used in wood stoves and fireplaces.

Cradle-to-cradle packaging: Packaging that is designed to be recycled, either by the consumer or through some natural process, unlike cradle-to-grave packaging, which anticipates no future use and is designed to be discarded.

Creosote: Specifically wood creosote, a by-product of the combustion of wood or paper and the subsequent cooling of the vapors.

Daylighting: The act of using natural light, as opposed to artificial light, to illuminate working and living spaces.

Deconstruction: The purposeful disassembling of a building, with recycling of materials as the object.

Desalination: The process for removing salt from seawater.

Dimethyl ether: Also known as DME and marketed as ethylene glycol, a colorless gas used as a propellant, a blowing agent for foam, a solvent, and a substance to create chemical reactions and extractions.

Direct current: Electricity that moves in one direction only, commonly used to charge batteries and found in most electronics applications. Direct current is the initial output of solar and wind-energy technologies.

DRAM: Dynamic random access memory, as it pertains to the storage capacity of a computer.

Earthship homes: Homes built within the concept of sustainability, often of earth-rammed tires or other waste products, featuring passive solar heating and other environmentally friendly energy resources.

EER: Energy-efficiency rating, a measure used to determine the energy efficiency of window air conditioning units.

Energy Star: A United States government program that sponsors energy-efficient products and provides information, documentation, and resources for energy-conscious consumers.

Energy vampires: Appliances or electronics that continue using reduced amounts of energy even when turned off. The loss, also known as phantom energy, is most conspicuous in computers, LCD displays, and televisions.

Engineered wood: Wood made by bonding strands, scraps, or fibers of wood with adhesives, producing veneers or composite lumber.

EPS expanded polystyrene: A rigid, cellular plastic, used for beverage coolers, packaging for electronics, small household appliances, and in insulation panels for building.

Eutrophication: A process whereby bodies of surface water receive excess nutrients in the form of fertilizers from runoff, resulting in oxygen depletion, loss of beneficial plants, and fish kills.

Fahrenheit: A measure of air temperature used in the United States and Jamaica.

Fluorescent: A form of lighting in which ultraviolet light is mediated inside a glass shield by a fluorescent material such as calcium fluoride, called a phosphor, and delivered as visible light.

Forest Stewardship Council: The Forest Stewardship Council (FSC) is a nonprofit foundation that certifies sustainably harvested wood, which is wood harvested using sustainable forestry techniques as opposed to the importation of wood obtained by illegal logging.

Formaldehyde: A colorless gas with the chemical symbol HCHO, commonly used in fertilizers, dyes, paper, clothing, and some cosmetics. It is also a disinfectant and a preservative and is used in embalming. Formaldehyde is a known carcinogen.

Fossil fuel: Dead organic matter that has been compressed by the heat and pressure of geologic events over hundreds of millions of years to form fuels such as coal, oil, and methane gas.

Fungicide: A chemical compound designed to destroy fungus. Natural fungicides include pyrethrins, sulfur, and copper. Man-made fungicides include Formec and Teremec.

Glacier: A large sheet of ice forming over a landmass as a result of compacted snow in areas where snowfall exceeds snowmelt. Antarctica contains glaciers, but the whole of the continent is a polar landmass.

Global warming: A gradual rise in earth's temperatures, most noticeable over the past 50 years and often attributed to the burning of fossil fuels.

GMO: Genetically modified organism, or an organism whose DNA has been mechanically altered to give it specific properties, of drought or disease resistance, for example.

Gray water: Water recycled from washing dishes or clothes or bathing, which can be used on lawns and ornamentals if sufficiently free from household chemicals or in which the chemicals are sufficiently diluted.

Green building: A new technique and focus in commercial and residential construction that aims for both energy efficiency and sustainability by achieving high insulative values, using passive solar techniques, and the use of recycled or sustainable building materials.

Green Pricing: The method utility companies use to determine the pass-through cost of technologically advanced forms of energy such as solar, wind and hydroelectric, which are called renewable energy.

Green roof: A roof that supports a layer of living plants.

Greenhouse gas: Gases such as nitrogen oxide, sulfur dioxide, and carbon dioxide, which trap heat in the earth's atmosphere, making life possible. In excess, these gases cause advanced warming, which can create a number of adverse effects on climate.

Greenwashing: A marketing technique used to label a corporation, or other entity, as environmentally responsible when the opposite is true.

Grubka stove: A highly efficient masonry stove design.

GSHP: Ground source heat pump.

Gulf dead zone: A large area in the Gulf of Mexico, off the Mississippi River inlet, that experiences annual and increasing eutrophication, or die-off of aquatic plants and fish.

Halogen bulb: An incandescent bulb or lamp that contains a tungsten filament sealed inside a transparent envelope filled with iodine or bromine, commonly used for automobile headlights or outdoor specialty lighting.

HEPA: High-efficiency particulate air cleaner. HEPA is a type of filter, not a brand name, and qualifying filters must capture at least 90 percent of airborne particles larger than 0.3 microns in diameter.

HVAC: An abbreviation for heating, ventilation, and air conditioning, used to describe a building's entire air-transfer system.

Hydroelectric energy: Also called hydropower, or electrical energy derived from moving water that passes over turbine blades, as in a dam.

Incandescent: A form of lighting in which electricity passes through a thin filament encased in glass to prevent rapid oxidation. Invented by Thomas A. Edison in 1879.

Indoor air quality: The presence or absence of indoor air particulates and volatile organic compounds (VOCs).

Infiltration: The ability of outside air to penetrate a building's envelope.

I.C.F.: Insulated concrete-formed houses, which are made by pouring cement into rigid foam insulation forms, not to be confused with Structural Insulating Panels (SIPs), which are made partially of wood.

Insulation: Any substance used inside or outside a building to prevent air infiltration.

Inverter: A device that transforms direct current to alternating current.

Kilowatt-hour, or kWh: The amount of energy produced, transmitted, distributed, and consumed in one hour, which equals 3,600,000 joules. Utilities use it to measure the average amount of power consumed by an entity in one hour. Rates vary among commercial, industrial, and residential entities.

Landfill: A place where items discarded from households, businesses, and industrial manufacturing are collected. Referred to in the vernacular as a "dump."

Latex paint: Paint manufactured using synthetic polymers rather than latex, which comes from the rubber tree. Latex paint is water soluble when wet but water-resistant when dry.

LCD: Liquid crystal display, a technique using two sheets of polarizing material with a liquid crystal solution between them that either blocks light or allows it to pass through, depending on ambient lighting conditions.

LED: Light-emitting diode, a form of electroluminescence that uses a treated, or "doped," semiconductor diode and the consequent polarity to create light by moving an electrical current in a forward (positive to negative) direction.

Linoleum: An environmentally friendly flooring material derived from linseed oil, which comes from the flax plant. When linseed oil is oxidized, it makes a solid compound.

Log homes: Homes made from peeled or unpeeled lengths of harvested trees, usually joined with saddle notches and chinked with grout.

Low-e window: Low-emissivity window, or low-emissivity glass, which is glass coated with a special substance designed to reflect or absorb infrared-radiation light, or heat energy.

Low-flow toilet: Also called a low-water toilet, a low-flow toilet is a toilet that uses less water than conventional toilets manufactured before 1994.

LPG: Liquefied petroleum gas, a hydrocarbon used to heat homes and power vehicles and increasingly as an aerosol agent and refrigerant.

MCS: Multiple chemical sensitivities, a symptom or syndrome of unknown cause resulting in multiple allergic reactions, which can range from annoying to life-threatening, also called Environmental Illness, appearing in 16 percent of the population and 33 percent of Gulf War Veterans.

MRSA: Methicillin resistant Staphylococcus aureus, a term used to describe a number of bacterial strains that are resistant to common antibiotics.

MSDS: Material safety data sheet, a publication that lists the contents of, and known hazards of, a chemical compound.

Mulch: A substance, usually organic and usually fragmented, used to hold moisture in the ground around plantings, not to be confused with compost, which can also be used as mulch.

Municipal solid waste: The contents of a landfill. Also see landfill.

NAHB: The National Association of Home Builders, which produces a publication called *HousingEconomics*, providing the latest housing forecasts, market trends, in-depth economic analysis, and archival data relating to the housing industry.

Negative ions: Negative ions are molecules that have lost an electrical charge and are produced in nature as a result of sunlight, radiation, moving air and water and other operations. Inside the home, they are generated by ionizing air cleaners.

Net metering: An energy policy applied to consumers, small businesses, and industrial operations wherein electricity generated by these entities is returned to the transmission and distribution system of a utility company.

Nitrogen oxide: A greenhouse gas produced by combustion at high temperatures, as in power plants and automobiles.

Nonrenewable resource: A resource which, once harvested, does not readily replicate itself or disappears from the ecology, as in wood from trees or fossil fuels.

Old-growth wood: Also referred to as old-growth lumber or antique wood. It is lumber or wood from a tree more than 100 years old, obtained through actual cutting of the tree or by salvage.

Organic gardening: Gardening without the use of fertilizers, pesticides, or fungicides.

Ozone pollution: Pollution of the earth's breathable atmosphere by excessive amounts of ozone, caused by excessive emissions from power plants, industrial operations, and automobiles or by the heating of chemicals. Not to be confused with the ozone layer, which rests above the atmosphere and protects the earth from radiation.

Papercrete: A substance made from a mix of paper and concrete used to form blocks, which are then used build the walls of houses.

Passive solar: Also called passive solar design, it refers to the orientation of a house or building designed to collect as much of the sun's warmth as possible to use as heating.

Peak oil: The point at which the use of oil or other fossil fuels begins to exceed the ability to produce it; a supply-and-demand scenario that predicts the need for alternative sources of energy.

Persistent Organic Pollutants: Also known as POPs, these are chemicals that persist in the environment for long, or sometimes indeterminable, periods of time, become distributed geographically by the actions of wind and water, accumulate in living tissue, and are toxic to all life forms.

Pesticide: Man-made chemical compounds designed to kill insects or reduce their ability to reproduce.

pH: A measure of soil's acidity or alkalinity.

Photovoltaic: See solar energy.

Pitch, roof: The inclination of a roof based on how many inches it rises per foot of length.

Polar landmass: Antarctica and Greenland are the two terrestrial polar landmasses. Each contains glaciers, or individual and identifiable masses of ice and snow.

Polybutyrate adipate terephthalate: Also known as PBAT, a biodegradable polyester.

Polyethylene terephthalate: Also called PET, a complex-molecule polymer resin, liquid when hot but solid when cooled, called a thermoplastic polymer. It is used in polyester and other fibers and also used to make food and beverage containers.

Polyvinyl chloride: Or PVC, another thermoplastic polymer that can be made more flexible by adding plasticizers, or phthalates. It can also be chlorinated.

Post-and-beam construction: Also known as timber-framing, the use of very large and/or very heavy timbers, or pieces of wood, to create the frame for a house or other structure, negating the need for 2-by-6 studs every 16 inches, as found in stick-built homes.

Post-consumer recycled content: Material salvaged after consumer use, such as wood, glass, and paper, and recycled into new consumer products with similar or different functions.

Programmable thermostat: A residential or commercial thermostat that can be preprogrammed to control the levels of heat, and the times delivered, inside a structure.

Psi: Pounds per square inch, a measure of the operating pressure or tensile strength of a material based on how much external pressure or weight it will accommodate and still perform within specifications.

Quad: One quadrillion Btus. One quad is equal to 10^{15} Btus.

Radon: Radon is a radioactive gas formed by the decay of radium naturally occurring in the earth, which escapes into foundations and basements and from there into buildings. Radon is a known carcinogen.

Rain garden: A declivity in the earth, natural or man-made, which accumulates and then evaporates water and, as a result, creates a specific ecosystem of plants and insects.

Rammed earth: A style of home building that uses earth, compressed under pressure, to form walls.

Renewable Energy Credits: Also called RECs, or Green Tags, are certificates that document a specific quantity of energy, commonly one megawatt-hour, that was generated using approved renewable energy resources such as wind, solar, or hydropower. These can be traded or used to provide funding for the expansion of renewable energy.

Renewable energy: Energy generated from renewable, i.e., persistent, sources such as the sun, wind, or water.

RESNET: The Residential Energy Services Network, a nonprofit organization dedicated to ensuring the success of the building energy performance certification industry, setting the standards of quality, and increasing the opportunity for ownership of high-performance buildings.

R-value: R is a measure of the thermal resistance, or nonconductiveness, of a substance. When applied to insulation, it is commonly calculated per inch of thickness.

SEER: Seasonal Energy Efficiency Ratios, a system used to calculate the energy efficiency of HVAC appliances.

Sick-building syndrome: Occurs when a building creates acute health complaints among its inhabitants, even though no specific cause can be isolated. Sick-building syndrome is attributed to a lack of adequate ventilation, a result of increasingly airtight building envelopes.

Solar energy: Or solar power, a method for using the sun's rays to generate electricity, not to be confused with a solar collector, which traps and stores heat for use in buildings or to heat water or swimming pools. A solar blanket for a pool is a simple example of a solar collector.

Stone construction: A method for building a house's walls and foundation from stone and mortar.

Straw bale: A method for building a house's walls from bales of straw, usually used in conjunction with timber framing.

Structural insulating panel: Also known as an SIP, it is made of insulating foam sandwiched between two layers of structural wood and is used to reduce the wood required for framing a house and to provide insulation.

Sulfur dioxide: A greenhouse gas generated primarily by the burning of fossil fuels.

Sustainability: The science and art of occupying the planet in such a way that the needs of the current inhabitants are met without compromising the needs of future inhabitants.

Tankless water heater: Also known as a demand, or on-demand, water heater, a water heating unit that responds to demand for hot water by heating it as needed at very high temperatures, as opposed to a "tank" water heater, which heats water continually.

Thermal conductivity: The ability of a medium to conduct heat over a specified time through a certain thickness in one direction, dependent on temperature gradients. Glass has a higher K-value than wood, according to a formula dictated by Fourier's Law.

Thermal mass: The ability of any material to store heat. Used in building, thermal mass refers both to the density and conductivity of a material.

Truss: A building component made of several triangular units joined at their axes by continuous beams, designed to resist tension and compression.

Ulraviolet radiation: Ultraviolet (UV) radiation is light whose wavelengths are shorter than visible light, making it both imperceptible to the human eye and greater in energy, which makes it useful in killing microorganisms. The earth is protected from most ultraviolet radiation by the ozone layer.

Utility grid: A system of transmission and distribution power lines, including substations, which utility companies use to transfer energy from the production site to the consumer. A person who is "on the grid," or "grid-tied," receives and/or returns power to this system of power lines, substations, and power plants. It is called a grid because it is based on a series of intersecting vertical and horizontal axes used to structure transmission and distribution. Being "off the grid" means a building is not connected to a power distribution line.

UV: See ultraviolet.

Volatile Organic Compounds: Also called VOCs, they are flammable substances that evaporate into the air, such as lighter fluid, gasoline, and paint thinner.

Whole house fan: A large, mechanical ventilation unit with a fan-blade diameter of 30 inches, usually situated in an attic next to a sidewall opening and which is used to cool the interior of a house by removing hot air. Not to be confused with an attic fan.

Wind energy: Electrical energy derived from a wind turbine and converted to household use with an inverter.

Xenoestrogens: Part of a heterogeneous group of man-made chemicals such as phthalates, as opposed to archiestrogens, which occur in nature. Xenoestrogens mimic the effect of natural estrogens and have documented ecological and health effects.

Xeriscaping: A gardening technique that uses minimal water and adaptable native vegetation to conserve water.

Zoned heating and cooling systems: Systems that allow the occupants of a building to control the temperatures of individual rooms. Not to be confused with programmable thermostats, which can vary heating and cooling needs only on the basis of the entire structure.

Bibliography

Books

Chiras, D. *The New Ecological Home*. Chelsea Green Publishing Company, White River, VT (2004).

Johnston, D. and K. Master. *Green Remodeling*. New Society Publishers, Canada (2004).

Roberts, J. *Good Green Homes*. Gibbs Smith Publishers, Layton, Utah (2003)

Venola, C. and K. Lerner. *Natural Remodeling for the Not-So-Green House*. Lark Books, New York, NY (2006).

Wilson, A.. *Your Green Home*. New Society Publishers, Canada (2006).

Web sites

Carmelo Ruiz Marrero, Biodiversity in Danger: The Genetic Contamination of Mexican Maize (online). (Cited Dedember 13, 2007). Americas Program (Silver City, NM: Interhemispheric Resource Center, June 2004). **http://americas.irc-online.org/pdf/articles/0406contam.pdf**

Climate.org, a Project of the Climate Institute, Impact of Climate Change on Human Health (online). (Cited November 27, 2007). Climate Institute, 2000-2003. **www.climate.org/topics/health.html**

Department of Fisheries and Wildlife, Pesticide Research Center and Institute of Environmental Toxicology, Natural Resources, Michigan State University, PCBs, DDE, DDT, and TCDD-EQ in Two Species of Albatross on Sand Island, Midway Atoll, North Pacific Ocean (online). (Cited December 16, 2007). Mindfully.org, June 1006. **www.mindfully.org/Heritage/Albatross-PCBs-DDE-DDT-Mar97.htm**

Electric Light & Power, U.S. Supreme Court upholds FERC rules on avoided costs, power interconnections with cogenerators (online). (Cited December 21, 2007). Electr. Light Power Journal Article, Volume 61, Issue 7. **www.osti.gov/energycitations/product.biblio.jsp?osti_id=5671729**

EPA Office of Compliance Assistance and Pollution Prevention, Junk Mail Reduction (online). (Cited December 15, 2007). EPA, Ohio division, July 2003. **www.epa.state.oh.us/opp/consumer/junkmail.html**

Hayden, Thomas. Trashing the Oceans (online). (Cited November 28, 2007). *U.S. News & World Report*, November 4, 2002. **www.mindfully.org/Plastic/Ocean/Trashing-Oceans-Plastic4nov02.htm**

Inter-Organization Programme for the Sound Management of Chemicals, Diethylene Glycol Dimethyl Ether (online). (Cited December 17, 2007). World Health Organization (INCHEM), 2002. **www.inchem.org/**

documents/cicads/cicads/cicad41.htm#8.3.1

National Academies, Colorado River Basin Water Management: Evaluating and Adjusting to Hydroclimatic Variability (online). (Cited December 13, 2007). National Academies, February 21, 2007. **www8.nationalacademies.org/onpinews/newsitem.aspx?RecordID=11857**

National Oceanic and Atmospheric Administration, Global Warming Frequently Asked Questions (online). (Cited November 22, 2007). NOAA Publications, 2007. **www.ncdc.noaa.gov/oa/climate/globalwarming.html#Q1**

Perspectives on Sustainability, Sustainability Defined, Ted Kesik P. Eng, Ph.D. (online). (Cited November 28, 2007). Agricultural Science Forum, January 2002. **www.canadianarchitect.com/asf/perspectives_sustainibility/index_frameset.htm**

Southern Oxidants Study, College of Forest Resources, N. Carolina State University, Southern Oxidants Study, Questions and Answers about Ozone Pollution (online). Cited December 22, 2007). SOS, June-July 1999. **www.esrl.noaa.gov/csd/SOS99/Ozone.pdf**

Tampa Bay Estuary Program, Nitrogen: Too Much of a Good Thing (online). (Cited December 20, 2007). TBEP, November 2003. **www.tbep.org/press/nitrogen.html**

Technology Review, Measuring the Polar Meltdown (online). (Cited November 27, 2007). MIT Publications, November-December 2007. **www.technologyreview.com/Energy/19504**

Union of Concerned Scientists, Are Energy Vampires in Your Home (online). (Cited December 12, 2007). UCS publication, May 15, 2006. **www.ucsusa.org/publications/greentips/energy-vampires.html**

U.S. Centers for Disease Control and Prevention, Four Pediatric Deaths from Community-Acquired Methicillin-Resistant Staphylococcus Aureus — Minnesota and North Dakota, 1997-1999 (online). (Cited December 23, 2007). U.S. CDC, August 20, 1999. **www.cdc.gov/mmwr/preview/mmwrhtml/mm4832a2.htm**

U.S. Centers for Disease Control and Prevention, Prostate Cancer: Can We Reduce Mortality and Preserve Quality of Life? (online). Cited December 16, 2007). CDC, 1997. **www.cdc.gov/CANCER/prostate/prospdf/proaag97.pdf**

U.S. Department of Energy, DOE Launches Change a Light, Change the World Campaign (online). (Cited November 30, 2007). DOE, October 3, 2007. **www.energy.gov/news/5552.htm**

U.S. Department of the Interior, Water Use in the United States (online). (Cited December 14, 2007). U.S. Geological Survey, August 28, 2006. **http://ga.water.usgs.gov/edu/wateruse.html**

U.S. Environmental Protection Agency, Energy and You (online). (Cited December 4, 2007). U.S. EPA, September 5, 2007. **www.epa.gov/cleanenergy/energynyou.htm**

Author Biography

The author was born in Colorado and spent much of her early life traveling around the Southwest with her father, Luverne Cartier, an engineer and building contractor who advocated the use of sustainable design before it was popular.

She has lived in a variety of places, including California, Arizona, New Mexico, and Chicago, and in a variety of homes, from adobe to earth-sheltered.

In 1982, her home in Minnesota won certification from the National Wildlife Federation as a National Wildlife Sanctuary. She has been published in *Face to Face Magazine*, *The Mother Earth News*, and the *East West Journal.*

Formerly a reporter for a California newspaper, she has also worked for a Chamber of Commerce, as the CFO for a utility district, and in the

Corporate Communications Department of Xcel Energy of Minnesota. She currently lives in Minnesota and devotes all her time to freelance writing.